MEGA PROJECT MANAGEMENT

The Secret of Successful Billion Dollar Project

Mega Project Management:

The Secret of Successful Billion Dollar Project

By Lim Guan Leng

Acknowledgement

I'm eternally grateful to my mum Madam Goh A.H. She taught me discipline, love, respect and so much more that had helped me succeed in life.

I would like to express my gratitude to all my previous bosses, especially Mr. Brian, who entrusted me to a few mega projects in Central Asia. These experiences widened my vision and allowed me to improve on my Project Management skills especially dealing with a multi-culture workforce. I learned to cope with engineering difficulties due to environmental or political factors.

I am thankful to my editor Tanha Emita for making important suggestions that enhanced this edition.

Lastly, I would like to say thanks to people who helped me along the way during the execution of the project overseas. We had overcome the difficulties and had made the projects a success and we are proud of what we had achieved.

Copyright © [2022] [Lim Guan Leng]

All rights reserved. No portion of this book may be reproduced in any form without permission from the author. For permission, Contact: alfred@brila.co.uk

Table of Content

- Global Mega Project – Operationalizing Coordination
- Culture differences between the East and the West
- Managing Global Project management – Complexity of Diversity
- Advantages and disadvantages of dealing with multi-cultural team
- Different Language leading to different expression in Communication
- Translation differences lead to misunderstanding
- Importance of Culture Awareness
- Persuading: Principles vs application

- Stakeholder Communication
- Trusting: Task based vs Relationship based
- Leading Across Cultures (Hierarchical vs. Egalitarian Leadership)
- Providing Feedbacks Across Cultures
- Benefits of Global Diversification
- Creating a Mega Project Culture

CHAPTER 1
Global Mega Project: Operationalizing Coordination

Project coordination includes all aspects of a project's planning, monitoring, and control, as well as everyone involved motivation. The project must be finished on schedule, on-budget, and to the desired standard of quality and performance. Coordination is a complex phenomenon with numerous facets. The two major goals of project coordination are to streamline the workflow of your tasks and to make sure that everyone is performing their responsibilities and efficiently contributing to project goals. Project coordinators typically assist project managers and support the team members as well.

Due to their inherent differences, projects are managed and coordinated in various ways. Marmgren and Ragnarsson (2001) defined three fictitious approaches: Weber, Rambo, and Gaia, which might be used to categorize various project management methods. Adler (1999) made a similar classification in a thesis based on related information. Weber is the abbreviation for the conventional project management methodology. The fundamental tenet is that difficult tasks may be managed and controlled by being divided into smaller and smaller bits and meticulously preparing each one. It is assumed that greater control is provided by more thorough preparation. The majority of project management literature lists this strategy as the best way to handle projects (Cleland and King 1988, Meredith & Mantel 1995, PMBoK 2000, Kerzner 2001, Milosevic 2003). This method frequently fails in large, dynamic projects like many of the modern mega-projects. Due to the rapid advancement of technology, it is practically impossible to construct a strict strategy during the project's start-up period. These projects may last a couple of years and should be state-of-the-art when they are delivered.

If project managers want to succeed in the present day, they must concentrate on multiculturalism. What is multiculturalism, exactly? Dealing with a group of internal workers from various racial and cultural origins might be necessary. It might also relate to managing a diverse team on an international project.

The most popular mega projects aim to get beyond the control issues that arise when results are supplied by progressively dispersed project teams rather than a vertically integrated company. Mega-projects are distinguished from conventional projects by their size, complexity, number of participants, and longevity. There are other projects with a multi-million euro budget, thirty partners, and a 20 year lifetime. These project-based alliances are examples of hybrid organizations, which incorporate aspects of networks and traditional hierarchical administration. A high degree of complexity is also present in many mega-projects due to a combination of joint organization and subcontracting of workflow components to legally separate partners. These factors together contribute to the high level of uncertainty and uncertainty that many mega-projects are characterized by. Due to the project's sheer magnitude or other factors that need reciprocal cooperation, the partners will still need to work together to convey complementing technology even if they are competitors. In reality, each partner wants to advance their own interests, wants to direct how tasks are completed, and is expected to support the alliance in one or more important strategic areas.

Alliances must acknowledge and encourage a dynamic viewpoint on issues that result from their hybrid nature, according to Child and Faulkner (1998: 41). Pitsis, Kornberger, and Clegg (2004) suggest a dynamic viewpoint on interorganizational interaction that takes power, cultural fragmentation, ambiguity, and complexity into account. Power, culture, contract, trust, and leadership are essential for an alliance's successful synthesis. Synthesis happens when organizational collaboration is successful between two or more organizations (Pitsis et al., 2004: 47). Therefore, power, organizational culture, fragmentation, and ambiguity should all be taken into account when examining the control-versus-commitment problem in mega-projects.

How commitment and control may be organized to achieve the goals of the mega-project is a crucial question. Cooperative activity organization can take on a variety of shapes (Child and Faulkner, 1998: 38). A dominant partner at one end of this spectrum sets up the massive projects in a hierarchical manner. The network model, which links cooperating parties through a variety of relationships, would be at the other extreme of the spectrum. In complex projects, Wijnen, Renes, and Storm (2001: 188) define three fundamental types of cooperation. The project manages content-related issues under the first category, which is the consultation or coordination model. While partners have a lot of authority to achieve project goals, project management has minimal power and authority. The matrix structure is the second type, and it is used by project management to coordinate context-related issues and start tasks. Partners carry on and oversee subsidiary ventures. The 'pure' project structure also affords partners minimal control. All actions are planned, carried out, and controlled by project management. Partners contribute resources, people, and knowledge. Each significant project must consult with the partners involved to determine the scope and emphasis of the control mechanisms in order to achieve its goals (Child and Faulkner, 1998: 187).

For collaboration in mega-projects, the point of control is a delicate subject. They may become the topic of ongoing political attention due to the tremendous media interest in their enormous budgets and the significant societal impact that these mega-project coalitions may have (Flyvbjerg et al., 2002). As a result, thorough financial audits and parliamentary control are probably in order. Regarding the scope of control, Three categories of alliance are distinguished by Child and Faulkner (1998: 188). One spouse controls decision-making in the first category. A shared management partnership falls under the second classification. Each partner actively participates in managing the alliance. The third category allows the alliance management to make decisions on its own. Project creation and execution should be centralized inside one project organization, suggests Flyvbjerg (presentation by Flyvbjerg in The Hague, September 2004). Construction companies, (sub)contractors, and operators can be held accountable upon request by the project organization, which may be public or private. Flyvbjerg contends that schedule delays, design flaws, and cost overruns are all the fault of the project organization and its management.

Organizations and project teams each have their own personalities, value systems, and operational procedures. If a project leader has a greater understanding of the concept of culture, they will be more successful in garnering support and guiding the project through the numerous organizational minefields. Project managers usually engage with a variety of cultural groups simultaneously. Project managers frequently engage with the core cultures of their own companies, the subcultures of other departments (such as R&D, marketing, and sales, or manufacturing, each with its own distinct "ways of doing things around here to succeed"), or the core cultures of external customers. Project success depends on being able to communicate in the local culture's language. By avoiding practices that go against the ideas and values of the client business, effective communication with the surrounding culture can aid in the development of plans and strategies that are more likely to be acknowledged and time-honored.

The lead culture of the organization must be in line with the project culture, which project leaders have many opportunity to intentionally develop and mold. This is a crucial step in creating a positive team environment and setting the foundation for the success of the project. A company with a strong culture has shared values and standards of behavior for its personnel, which should aid them in achieving their objectives. When workers are able to finish the responsibilities given to them by the company, they can receive praise for their work and experience job satisfaction.

This book departs from conventional project management by concentrating on cultural and regional variations and affects on a particular project activity. It also discusses the special influences on major significant critical projects (also called "megaprojects"). References to pertinent methodology, theories, and findings from other fields are drawn from micro-behaviors impacted by regional cultures, socio-cultural theories, cultural synergy processes, hybrid institutions, physioeconomics, and various contractual procedures.

Utilizing cross-cultural project teams effectively can increase the likelihood of project success and strengthen the competitive position of the firm by supplying a source of creative thinking and different expertise. However, in today's multicultural, international business world, cultural differences and associated disputes might obstruct the successful completion of projects if they are not handled with care. Project managers should be culturally aware and encourage innovation, respect, and enthusiasm through flexible leadership in order to accomplish project goals and prevent cultural misunderstandings. The most popular and widely accepted theories of cultural differences are discussed in this chapter along with the findings of additional research and project management examples used to demonstrate them. These theories consider interpersonal relationships, motivational orientation, how one defines oneself and others, and how one feels about time, risk, control, context, and the environment. The findings of a study comparing divergent and convergent thinking using associative group analysis are presented in this chapter, along with additional project management implications. We talk about the training and motivation of multicultural project teams as well as any consequences for project management that are pertinent. We give particular instances of multicultural initiatives' successes and failures and link project results to cultural variances. This chapter concludes that culturally competent leadership, successful cross-cultural communication, respect for one another, and reconciliation are all necessary for multicultural project management to be successful. It is destined to fail without them.

CHAPTER 2

Cultural Differences Between the East and the West

"When working in a multicultural team or in a different country, it is imperative to understand how different cultures communicate in the workplace and how to adapt and embrace those differences."

While the Western world includes North and South America, Europe, Australia, and New Zealand, the Eastern world includes nations in Asia and the Middle East. Based on culture, East and West may differ greatly. The key indicators of these inequalities are people's actions and attitudes. But it is only possible to compare the two civilizations on a broad scale because the terms "east" and "west" relate to a variety of nations and cultures dispersed throughout the world. People in the east are more traditional and conservative than people in the west, which is the main distinction between eastern and western cultures.

One of the most significant distinctions between western and eastern cultures when they are compared is the degree of liberalism in western countries relative to those in the east. Western culture encourages people to be more critical and open-minded. They talk about topics that are taboo in eastern cultures, and they are free to express their feelings, even rage, if they feel it is appropriate. Eastern civilizations wouldn't tolerate this kind of conduct. Instead of acting aggressively, people prefer to address challenging situations with grace and tact.

Another difference between eastern and western cultures is that people from the West have more freedom and flexibility to make decisions on their own, as opposed to people from the East, where families tend to make decisions more jointly. Thirdly, since love is regarded as the only method that people get into marriage, arranged marriages are not a prominent aspect of western societies.

The western nations are frequently referred to as "The Global North" and the eastern nations as "The Global South."

It's interesting to note that the majority is made up of individuals from eastern cultures and countries. In the world, only one out of every five people is from a western culture; the other four all come from eastern cultures. Thus, the odd minority is made up of Westerners.

Comparing western and eastern cultures reveals that western education emphasizes innovation and gives students the freedom to reach their full potential. Achievement in Eastern education is correlated with adversity and diligence. This suggests that if you work hard enough, you can do anything. Because they put forth more effort than western children, students from eastern cultures frequently achieve academic success in western educational settings.

Western culture encourages students to actively engage in discussions and pose questions. Eastern societies, where what the instructor says is always correct, do not hold this view as strongly. Western cultures go above and beyond to integrate kids who are labeled as having special needs. Together with other pupils, they sit in classrooms. In eastern cultures, where children with exceptional needs are taught separately, this doesn't happen very often.

Overall, there aren't many similarities between eastern and western civilizations at this moment, but each one should be honored for what it stands for and what it has accomplished.

The most significant cultural divides between the East and the West are listed here.

1. Business correspondence

Depending on how much weight each culture places on indirect or direct communication, you may distinguish between distinct intercultural communication styles, according to renowned anthropologist Edward Hall. Edward Hall created the idea of high-context and low-context cultures and authored a number of books on navigating cultural differences.

cultures without context

Low-context cultures, including those in Germany, the US, and Australia, focus on direct communication and the use of concrete language to make their points; more details are stated out and defined in a message in these cultures.

"They think what they say and say what they think,"

Low-context cultures are often seen to have roots in Western Europe. The majority of low-context cultures encourage open communication and consider it acceptable to voice anger, annoyance, or dissatisfaction.

context-rich cultures

High-context cultures, like those in China and Japan, place more emphasis on nonverbal, indirect communication. These cultures will favor preserving a sense of general peace and avoiding conflict at all costs. In reality, since plain conversation implies "losing face," it is strongly discouraged in Asian societies, which are based on a Confucian ideal of social harmony and rigid hierarchy.

The heart of Chinese culture is the idea of "losing face." When you lose face, it indicates that the social order's trust in you has been compromised. Your reputation has suffered, and you have lost influence. Therefore, it is unacceptable to criticize a coworker in front of other team members in China. Instead, criticism is reserved for private conversations and frequently conveyed via a third party.

2. Business connections

Business ties start and end rapidly in nations like the US and Australia because of their low-context cultures. Procedures and paying attention to the objective are essential for productivity. For instance, the majority of American executives can start a professional connection with a European or American executive through one or more phone calls or meetings.

Higher context cultures

Developing relationships is crucial while conducting business in regions like Asia. However, relationships take longer to form in Asia, so the secret is to get introduced via friends, coworkers, or classmates. Asians place a high value on interpersonal connections and make an effort to establish a personal bond with everybody they conduct business with.

(1) The value of hierarchy

Depending on the culture you were up in, the director, CEO, or other leader is viewed differently in every firm. Unexpected misconceptions might result from variances in how leadership approaches are perceived across cultures. Power and power distance are viewed differently in Eastern and Western cultures. While Western cultures are more egalitarian, Eastern cultures frequently have a more hierarchical structure.

"Western cultures emphasize individuality and frequently support those who are goal-oriented and individualistic."

Employers in the West want their staff members to take initiative and demonstrate their own unique skills. On the other hand, an Asian leader will place more emphasis on group accomplishments and will anticipate that his or her team members will respect the organizational hierarchy by being devoted to it and cooperating with one another.

2. Punctuality

Both the East and the West's corporate cultures place a high value on punctuality for events linked to work. Nobody like waiting around for other people to show up, and being late is typically seen as impolite and unpleasant. There are a few potential issues, though, that you should be aware of.

Even between Westerners of various origins, local norms can be difficult in the West, especially in nations like Spain. Although everyone makes an effort to appear on time and begin meetings promptly, working flexible hours is becoming more and more accepted. There is typically nothing to be concerned about in this situation as long as the work is completed and meetings are attended. Having said that, always make an effort to come on time, but remember that people sometimes arrive a little late in various cultures.

Arriving on time is crucial, especially in eastern nations like Japan. In fact, Japanese railways employ workers to provide 'train delay certifications,' which will indicate the passenger's late was due to circumstances beyond their control and prevent them from getting into problems with superiors at work. The conductor will also convey his or her apology over the intercom. 3.

The simple act of asking a question might be seen differently by both the individual asking the inquiry and others watching the exchange. When completing your internship in an Eastern or Western nation, you will quickly come to understand the distinctions.

Employers in Western nations value employees who ask inquiries because it demonstrates initiative and initiative. Employers reward candidates that ask questions because it demonstrates their interest in learning and desire to contribute to the company.

On the other side, as was already noted, Eastern nations place more value on being courteous and refraining from addressing disagreements, viewpoints, or defects in public. Because they would have to explain their stance on a particular subject, bosses would interpret questioning as threatening.

Examples of failures in business brought on by cultural differences:

Understanding cross-cultural variations is vital for success when organizations decide to conduct operations with partners or customers from different cultures. The examples that follow can help illustrate how even a little communication or cultural misunderstanding can harm a business relationship.

the Heineken brand

The flags of the nations that qualified for the 1994 World Cup of Football were imprinted under the bottle caps of Heineken beers as part of an unique marketing campaign. The Saudi Arabian flag, which features a sacred text, was one of the flags present. As a result, many Muslims were incensed that a sacred verse would now be connected to alcoholic drinks. The brewer was forced to stop its promotion as a result of a straightforward cultural error.

Example of eBay

Trust is a value that is highly valued in Asia, hence an entirely transactional business strategy for online transactions is insufficient.

For online transactions to foster trust between the buyer and the seller, a successful communication option is a requirement. However, the eBay platform in Asia lacked the resources to let customers find online vendors and get in touch with them right away to discuss potential deals. eBay misunderstood the differences between US and Asian cultural norms and values.

CHAPTER 3

Managing Global Project management – Complexity of diversity

In the twenty-first century, transformational undertakings almost invariably require various team structures. Applying efficient team management techniques to various groups at the appropriate moment is a difficult task in and of itself. Effective collaboration, communication, and coordination, adaptive team leadership, optimal team structure, the ideal team composition, a disciplined culture, co-location of core team leaders, and patience to guide the groups as they develop from a collection of individuals into a good team and finally into a great team are just a few of the factors that come together to form successful teams.

It is not uncommon for project teams to include sponsors, consumers, architects, and developers dispersed throughout the globe since initiatives involving major change in how business is performed are almost going to involve complicated team structures. Constantly flying around the world to meet with team members in person is too expensive and simply too exhausting. We need to develop new approaches to managing complicated teams, combining robust virtual interactions with in-person meetings in order to fully benefit from big improvements to maximize business and technology.

Businesses are being forced by 21st-century needs to abandon conventional "command and control" management structures and develop cutting-edge methods for team composition by interacting with both the virtual and real worlds. Companies are creating innovative yet complex organizational communities in order to stay competitive. These relationships might be win-win arrangements with important political organizations, customer networks, regulatory bodies, and yes, even rival businesses. Organizations are responding to the demands of extraordinary change, global competitiveness, time-to-market compression, fast evolving technologies, and rising commercial and technological complexity through these creative alliances, which take the form of both physical and virtual models.

The difficulties of managing teams are significantly exacerbated by geographical diversity and reliance on technology for communication and collaboration. Applying the proper team management strategies to various stakeholders at the appropriate moment is a difficult task. The responsibility for project leadership shifts from project and requirement management to team leadership and group development.

We will first examine the kind of challenges that arise from managing multi-methodologies, complex teams with varying cultural norms, and complex contractual arrangements. These challenges include:

• Uncertainty in interactions; difficulties integrating

The utilization of complex team management strategies is then examined, along with creating a workplace that is flexible, innovative, and creative. The following topics will be covered: • Leveraging team potential; • Becoming a team leader; • Using tools and approaches for team cooperation, communication, and coordination.

Managing complex teams with diverse cultural norms, intricate contractual agreements, and varied approaches involves several complications. Here, we only look at a few-

As Complex Adaptive Systems, Teams

Human behavior is complicated because people constantly respond to their surroundings, making it hard to predict human behavior, as complexity science teaches us. Teams are complex adaptive systems functioning within a larger program, and the program itself is a complex adaptive system working within a complex adaptive organization that is attempting to succeed in a complex adaptive global economy. You cannot foresee how your team members will respond to one another, the project's requirements, or their position within the program and the larger company as the leader of a new, complex project or program. Therefore, complex team leadership is quite difficult. Since you are now managing via teams, stop thinking of yourself as a project or program manager and start honing your team leadership skills. You are a team leader, not a project manager, when overseeing a complicated project.

Uncertainty in Interaction

It seems at first that a group of individuals who have previously collaborated will soon develop into a high-performing team. They might also bring prejudices or resentments toward one another to the new team along with their baggage. Team members who haven't worked together before are more likely to be cautious until they get to know each other, the team dynamics, the task at hand, and their expected roles and responsibilities. This idea, known as "interactional uncertainty," acknowledges that when there is ambiguity in a connection, the participants will often tend to keep certain information to themselves and weigh the pros and cons of disclosing particular information. The leadership team for the project must mentor participants through the unavoidable early phases of team development to "interactional certainty," which develops into trust. Team members can then concentrate their efforts on productive relationships. It can be exceedingly difficult to create a trustworthy atmosphere, develop "interactional assurance," and hence encourage trustworthy relationships when working in a virtual environment.

Integration Issues

Working with numerous diverse teams nearly always results in integration problems, making it challenging to combine interdependent solution components that were developed and built by various teams. Teams frequently employ unconventional methods, techniques, and equipment, which yields work products of variable consistency and quality. Finally, flaws in a variety of project management strategies, such as risk management and complexity management, can have unintended repercussions that necessitate further work to fix.

Project managers must harness the power of teams, develop their team-leading skills, and become proficient with cutting-edge technology for collaboration, communication, and coordination if they are to successfully manage complicated layers of teams.

Teams are a vital resource that are utilized to boost performance in all different types of companies. However, business leaders today frequently miss opportunities to fully realize their potential because they conflate teams with teamwork, empowerment, or participative management. Without comprehending and utilizing the strength and knowledge of teams, we will be unable to address the problems of the twenty-first century, including business transformation, innovation, and global competition.

Effective complex project managers recognize the strength of teams. Success tales are numerous: By using teams as a strategic advantage, Motorola outperformed the Japanese in the race to control the cell phone industry. Similarly, 3M employs teams to achieve its objective of producing half of annual sales from innovations over the previous five years. The U.S. Navy Seals, tiger teams formed to carry out specific missions or tackle challenging problems, paramedic teams, fire fighter teams, surgical teams, symphony orchestras, and professional sports teams are just a few examples of high-performing teams that surround us. These teams are a convincing example of the strength of teams since they consistently show off their successes, perceptions, and zeal. Clearly, if we want to complete challenging projects, we must learn how to create, nurture, and sustain high-performing teams. Teams are effective. Utilize your teams to produce outstanding achievements!

Understanding the advantages of teams and learning how to enhance team performance through member development, team building, and rewarding team accomplishments are goals of effective complex project managers. Complex project managers cannot overlook the strength and intelligence of teams because they are the fundamental components of effective organizational performance.

Outstanding team leadership yields outstanding outcomes. How therefore do we prepare ourselves to excel as team leaders? Experience, team building and nurturing, excellent team makeup, and an ideal team structure are some of the "must haves."

Experience cannot be replaced.

People, not technology or science, are to blame when projects fail. Just as teams differ from operational work groups, team leadership is fundamentally different from traditional management. The complex project manager works via others; these "others" are responsible for project management. Leadership in a team is more of an art than a science, and it takes experience, try, and error to get it right. The ability to communicate effectively, solve problems, resolve conflicts, and possess other so-called "soft skills" is crucial. The capacity to develop relationships rather than holding a position of authority inside the organizational structure is what gives leaders of complex initiatives their power and influence. These leaders must be knowledgeable, powerful, connected, highly regarded, and even indispensable.

Discover how to create and maintain your team.

The dynamics of team development and how teams function must be understood by complex team leaders. Complex team leaders have specialized abilities that they utilize to create and maintain high performance. Without the proper mentality, mentoring, and training, traditional managers and technical experts may not always make good team leaders. Develop your team leadership abilities with a focus on transforming your team members into a cohesive unit with shared values, beliefs, and a morally sound cultural base. The best teams work together and alternate in the leadership position according to the specific requirements of the project at hand. The situational team leader is aware that different leadership philosophies are acceptable at certain team stages.

Get the "right thing" on your team by doing careful recruitment.

The most crucial choice you will ever make is choosing the ideal teammates for your team. Recruit team members that are passionate, strategic thinkers who thrive in a demanding, collaborative environment in addition to those who possess the necessary knowledge and abilities. According to conventional wisdom, we should first decide what needs to be done before choosing the right individual who has the necessary knowledge and abilities to complete it. However, Jim Collins clearly states: first who, then what, in his book Good to Great. Collins and his research team discovered that great companies actually did the exact opposite: They first chose the people who had the "right stuff" and then collaboratively set their course. This is in contrast to the traditional method of establishing a direction, a vision, and a strategy for your project and then getting people committed and aligned.

Create the ideal team structure.

Structure is crucial! As recommended by experts like Jim Highsmith and Jim Collins, typical modern team structures include:

a "hub" team or core group. Both hierarchical and network structures can be seen in this arrangement. Several customer teams, numerous feature teams, an architecture team, a verification and validation team, and a project management team are frequently included in this paradigm. Teams might be virtual, physically present, or any combination of the two.

expansions of self-organization. The organizational structure changes from a team framework to a project framework, where numerous teams work together, as the number of teams inside the project increases. Getting the correct leaders is the first step in developing a self-organizing team framework. The next steps are to communicate the task breakdown and integration techniques, encourage team engagement and information sharing, and frame project-wide decision-making. Obviously, complexity rises as more teams are formed. Teams must completely comprehend their borders and interdependencies in order to manage interteam dependencies.

a disciplined and empowered culture. When working in this structure, teams must exhibit the following behaviors: (1) accept responsibility for team outcomes; (2) collaborate with other teams; (3) adhere to the project organization framework; and (4) strike a balance between project and team goals.

The following practices should be taken into consideration for efficient team collaboration, communication, and coordination of complex team structures:

• A standardized approach

• Cooperative planning and decision-making; • Modern technologies for collaboration

A consistent approach promotes discipline and makes communication easier.

It helps much to eliminate unidentified cross-team dependencies when implementing a common methodology for complicated projects while encouraging each team to modify it as necessary. But a word of warning: Avoid overburdening the different teams with standards, but do insist on those that are required to manage dependencies between teams and to present a realistic image of the whole project. Make that everyone is using the same collaboration tools, methods, and procedures.

Planning and decision-making jointly fosters dedication.

Include every key team member in the project planning process, and frequently solicit input to help the team perform better. Face-to-face working sessions during planning meetings are indispensible, particularly when it comes to brainstorming, innovation, evaluating the viability of potential solutions, scoping, scheduling, identifying risks and dependencies, and performing crucial control-gate reviews. Make sure to allocate enough time and money in your project budget to bring the core team members together for these crucial meetings. Be adamant about establishing decision checkpoints that include the entire core project team at pivotal moments.

Modern collaboration technologies make consensus possible.

Best-in-class, secure software tools for document sharing and collaboration Professional service automation (PSA), which is intended to optimize service engagements, and enterprise project management (EPM) tool suites, which are used to manage numerous projects, are the two main categories of collaboration tools available.

In order to make your team members feel more bonded and connected, give them access to personal communication and telecommunications tools. Inform your project sponsor about the importance of collaboration, emphasizing the need to manage the cross-project interdependencies that are known at the start of the project as well as those that will emerge along the way, if these tools are an unconventional expense item for projects in your organizational culture. Try out different groups and social networks as well. Sites like MySpace and YouTube have benefited greatly from this computer-mediated communication, which has led to huge user bases and billion-dollar purchases of the software and the communities by big businesses.

There you have it: the complex, the good, and the bad of teams in the twenty-first century. A strong squad cannot be stopped. Great teams, however, do not develop by chance. A team of exceptional individuals must put in a lot of effort, preparation, and focused effort to become a great team. Due to the involvement of numerous sizable, geographically distributed, and culturally diverse teams in complicated projects, the effort is heightened. Complex project managers transition from being project managers to being team leaders. Large, long-term initiatives require both traditional and adaptive approaches to be effective.

CHAPTER 4
Advantages and Disadvantages of Dealing with Multi-cultural Team

There is no question that multiculturalism will have numerous benefits. The repetitive modern lifestyle is given color by multiculturalism, which also makes cities exciting and colorful. People arriving from diverse nations, each with their own customs and cultural history, will provide the locals with new, unusual experiences. Apart from Christmas, there are bright festivals and events virtually every week in Auckland, including the South Pacific Pasifika Festival, the Indian Diwali Festival, the Irish St. Patrick's Day celebration, and the Chinese Spring Festival. The citizens of Auckland would hardly ever feel bored because they love the cuisine, music, and dancing from various nations. All of these things will enhance the city's appeal and draw more tourists. Therefore, multiculturalism is beneficial to the economy because it will enhance employment, income, and local government revenue. Cities with a diverse population, like New York, Singapore, London, and Sydney, are excellent examples. Beyond this, coexistence fosters learning and mutual understanding among people of various ethnic backgrounds; more mutual understanding may be extremely helpful in eradicating systemic racism and prejudice.

Additionally, when people interact with people from different cultural backgrounds, they may realize that the existing elements of the world can be viewed in a variety of ways, which eventually fosters openness and inspires creativity. They may also combine different cultures to create new, original cuisine, architecture, music, and cultures. The best evidence of the benefits of multiculturalism comes from the people themselves, who, according to a survey, "support a robust multicultural society, with 89% saying that it is a good thing for society to be made up of individuals from diverse races, religions, and cultures."

Despite being few, multiculturalism's negative repercussions are nonetheless worth mentioning. First, due to factors like culture, discrimination, injustice, inequality, and religious convictions, there is a possibility of social conflict. The recent refugee riots in Europe, which resulted in many casualties and social unrest, would be one of the typical examples. In 2016, there were about 3,500 attacks against refugee camps or other locations housing refugees in Germany (Larsson, 2017 page 1). Second, if the influence of the dominant culture is too powerful, weak minority groups may unintentionally or consciously lose their original ethnic and cultural identity or way of life. Thirdly, it might be difficult for someone who belongs to a minority group to adjust to a new environment that is occasionally unwelcoming; as a result, they are likely to be on the periphery of society as a whole. Additionally, the initial academic success of their children is hampered by the unfamiliar language environment.

Flow of International Labor

Multiculturalism is mostly caused by the influx of international workers.

The total number of international labor migrants is steadily rising as a result of intense globalization. Because of their long-term coexistence and the fact that they bring their working skills, conventions, and religious convictions to a new country, diversity is unavoidable. Using Auckland as an illustration, local labor scarcity becomes a significant issue as the city's economy continues to grow steadily. According to a report, 32,000 additional workers would be needed in the infrastructure and construction sectors over the coming years. Therefore, the government must let firms to hire qualified overseas workers on talent visas from a variety of nations. The top five source countries for the 8,668 Essential

Skill workers in Auckland that were authorized with an offer of employment in 2014/2015 are as follows: Philippines 9%, India 21%, China 10%, Fiji 10%, UK 9%, and India. It demonstrates the ethnic and cultural diversity of the foreign workers imported into Auckland. As more and more foreign laborers arrived in Auckland, the city became more and more cosmopolitan.

Education

International students are a common cause of multiculturalism in addition to global labor migration. Studying abroad is becoming increasingly popular across the board as a result of the globalization of the economy. Students are keen to have the opportunity to experience exoticness and high-quality education in other nations. "There were around 5.1 million internationally mobile students in

2016." In addition to gaining academic information, international students are also influenced by local culture while also showcasing it to locals. One of the most desirable locations for international students is London, the capital and largest metropolis of the United Kingdom. London is the #1 destination for international students wishing to study abroad, claims the website. Every year, more than 100,000 international students from more than 200 different countries attend school in London. Following are the top 5 nations of origin: China, Malaysia, the United States, Hong Kong, and India. "International students are transforming London into a multicultural, energetic metropolis." As more and more foreign students arrived in London, the city became more multicultural.

Refugees

Refugees contribute much to multiculturalism as well. People who have been compelled to flee their country due to conflict or other humanitarian issues are referred to as refugees. They are a sad group, and the most of them are in extremely precarious financial situations. They must relocate to other nations and then adopt their culture and customs. The staggering figure is "there are 25.4 million refugees in the world." The majority of refugees from the West Country have been taken in by Germany. In 2017, 222,683 migrants entered Germany; the top 5 countries of origin are Syria, Iraq, Afghanistan, Eritrea, and Iran. A large number of these refugees resided in Berlin. Some of them reside in refugee shelters, while others share housing with locals. Berlin is a city with a diverse population of 3,388,434 people. The Top 5 citizenship origins are as follows: Poland 1%, Turkey 2.9%, other Europe 3.3%, Germany

88.6%, Asia 1.7%. Regarding religion, Protestants made up 19.2%, Roman Catholics 8.9%, and others 71.9%. Refugee immigration raised its level of multiculturalism. As migrants from many nations continue to arrive in Berlin, the city has become increasingly cosmopolitan.

The workforce of the twenty-first century can originate from everywhere, including the Midwest, Mumbai, Nigeria, Nova Scotia, and New York. It might be difficult to integrate a diverse workforce into a productive team, but success has many advantages. Businesses that are multi-cultural frequently have an advantage when serving customers from other cultures. The internal operations of your business could benefit from diversity as well.

Advantage: Diverse viewpoints

People from various cultures have various perspectives on the world. They learn various ways to complete tasks and are raised with various priorities, conceptions of what is achievable and appropriate. That is a significant benefit when you require some original ideas for your most recent endeavor.

Different Collaboration Styles are a Disadvantage

However, reaping the rewards of multiculturalism requires conscious effort. The ways in which different cultures collaborate are highly different. Building team consensus before proceeding is highly valued in many Asian and Central American cultures. The emphasis on individual thought and action is greater in American society. Questioning your superior or breaking bad news to him is a huge no-no in some cultures. To gain from multiple viewpoints, you must balance the various styles.

Advantage: Learning and Growth

All of your employees will gain if you manage your multicultural team effectively. They will gain knowledge of many cultures and acquire skills for bridging cultural divides. That might help them professionally. They might also get a chance to put any foreign languages they know to use, which is crucial for maintaining fluency.

Challenges with paperwork are a drawback.

The paperwork involved in creating a multicultural workforce is a drawback. The United States has stringent visa procedures for hiring personnel from abroad. The laws have changed over time and from one country to another. This shouldn't deter you, but it will undoubtedly need more work than simply hiring an Indianapolis resident.

Advantage: Experience with a Variety of Markets

Your company can grow into demographic and geographic areas that it might not have otherwise by hiring a multicultural team. For instance, employees from different cultures are intimately familiar with the cuisine that people like, the music that they listen to, and even the types and shades of clothing that they find most appropriate. This information can be used wisely to advertise your goods to new buyer demographics.

A drawback is the steep learning curve.

There are many difficulties when working with clients or suppliers from different cultures or nations. Many nations around the world have distinct handshake customs than the United States. For instance, in Turkey, you hold the other person's hand for a considerable amount of time while applying a gentle, not hard, squeeze. Having workers who are accustomed to different cultures allows you to quickly get "how to" guidance on intercultural communication. Additionally, they might be able to communicate with customers in Spanish, German, or Japanese.

The drawback is that similar issues can arise at work in a multicultural setting. All of your staff may need some coaching and time to become familiar with one another's traditions and nonverbal cues.

CHAPTER 5

Different Language leading to different expression in Communication

While the exchange of messages or information between people, whether verbally or nonverbally, is referred to as communication. Language, on the other hand, is a form of human communication or the mechanism through which two individuals engage in conversation.

It is utilized in a specific area or community for the purpose of verbally communicating with one another.

While the exchange of messages or information between people, whether verbally or nonverbally, is referred to as communication. Language, on the other hand, is a form of human communication or the mechanism through which two individuals engage in conversation. It is utilized in a specific area or community for the purpose of verbally communicating with one another.

Language is an essential component of communication. In fact, every living thing on earth speaks their own language. Because of how closely these two concepts are related, individuals frequently confuse the two and use them interchangeably. However, there is actually a subtle border between language and communication.

Language is said to as a tool that aids in the communication of thoughts and feelings between individuals. Through arbitrary created symbols or noises, such as words (spoken or written), signs, sounds, gestures, posture, etc., that carry a specific meaning, it is a person's way of expressing what they feel or think.

The only means of communication between two people that allows them to express their thoughts, ideas, opinions, and feelings to one another is language. Its goal is to clarify abstract and complicated ideas without creating confusion. People who live in different places or are a part of distinct communities use different languages as a means of communication.

The act of exchanging concepts, knowledge, or messages from one person or location to another using mutually understood words or signals is referred to as communication. Because it is a fundamental method of how organizational members interact with one another, communication is essential for the organization. It may move in a number of directions, including upward, downhill, horizontal, and diagonal.

As a pervasive process, communication is required at all organizational levels and in all types. It is a two-way activity made up of seven main components: the sender, the message, the channel, the receiver, the decoder, and the feedback. Receiving response during the communication process is just as crucial as conveying the message since only then will the process be finished. There are two ways to communicate, which are as follows:

- Communication in Form

Interpersonal Communication

The following points detail the distinctions between language and communication:

1. Language refers to the system of communication that depends on the verbal or non-verbal codes that are used to communicate information. Communication is the process by which two or more individuals exchange messages or information.

2. While communication is the act of passing messages from one person to another, a language is a tool used in communication.

3. Language emphasizes words, signs, and symbols. The message is emphasized in communication.

4. Language was limited to auditory means until written words were created. However, it can also happen in other sensory channels, such as the visual and tactile. Conversely, communication happens through every sensory channel.

5. The fundamentals of communication remain constant. On the other hand, the language's vocabulary is updated every day with new terms.

Language always refers to things outside of itself: Languages have specific meanings that reflect the culture of their respective social groups. Interacting with a language entails doing so with the culture that serves as its foundation. Because of their close relationship, we could not comprehend a culture without having direct access to its language.

The culture of a given social group is shown by the language spoken by that group. Therefore, learning a language involves not only learning the alphabet, the meaning of words, the rules of grammar, and the order in which they are used, but also learning about social behavior and cultural practices. Therefore, unambiguous references to the culture, the whole from which the specific language is taken, should always be made in language instruction.

Since many of our communications are sent through paralanguage, human communication is complicated. If the larger cultural context is not taken into consideration, contact with people from other civilizations or ethnic groups is subject to the risk of misunderstanding because these auxiliary communication approaches are culture-specific.

Growing up in a certain society, we inadvertently learn how to utilize gestures, looks, minor alterations in tone of voice, and other auxiliary communication tools to modify or accentuate what we say and do. Over a long period of time, primarily through observation and imitation, we learn these culturally specific techniques.

Body language, also known as Kinesics or the language of movements, expressions, and postures, is the most evident type of paralanguage. The tone and personality of a voice, however, can also change the meaning of words.

Culture is language, and language is culture.

The link between language and culture is intricate and analogous. Language and culture are intricately entwined (they have evolved together, influencing one another in the process, ultimately shaping what it means to be human). "Culture, then, began when speech was present, and from that point on, the enrichment of either entails the further growth of the other," observed A.L. Krober (1923) in this regard.

Cultural manifestations are acts of communication that are adopted by certain speaking communities, assuming culture is a result of human contact. The entirety of the messages we exchange with one another when speaking a certain language, in the words of Rossi Landi (1973), "constitutes a speech community, that is, the entire society understood from the point of speaking." He goes on to say that all children learn their native tongues from their communities, and that while they do so, they also acquire cultural knowledge and grow cognitively.

Language and culture both communicate through one another, and vice versa: According to Michael Silverstein, culture's communicative power not only serves to portray elements of reality but also to link one setting with another. In other words, communication is a technique of bringing beliefs, feelings, and identities into the current context as well as the usage of symbols that "stand for" thoughts, feelings, identities, or events.

The linguistic relativity principle states that the language we choose to describe the world has a direct impact on how we perceive it. "The language habits of the group are, to a considerable part, unintentionally used to build the real world. Never are two languages identical enough to express the same social reality. Different societies live in diverse worlds that are not just the same with a different label on them (Edward Sapir, 1929). As a result, speaking assumes a culture, and understanding a culture is similar to understanding a language. Culture and language share the same mental realities. The world is represented and interpreted via cultural artifacts, which must be shared in order to be experienced.

The issue arises from cross-cultural encounters, or when the message sender and message receiver come from different cultural backgrounds. Intercultural communication is essential for anyone who wants to get along with and understand people whose ideas and origins may be very different from their own. Contact between cultures is growing, and this is true for everyone.

Language is used to relate to things outside of itself and to mark cultural identity, especially when a speaker utilizes it to clarify their objectives. The culture of a given social group is shown by the language spoken by that group. As a result of the interconnectedness of language and cultural learning, we might assume that language learning is cultural learning and that language teaching is cultural teaching.

A group of people's common attitudes, beliefs, social norms, fundamental presumptions, and values are collectively referred to as their culture. This culture shapes each member's behavior as well as how they perceive the meanings of other people's behavior. Language also serves as a vehicle for the expression and embodiment of other occurrences. As a result of their socialization into that community, people of a given society express the values, ideas, and meanings that they all share. Proper names that represent those objects show that language also relates to culturally specific objects. According to Byran, "a loaf of bread" in British use conjures up a particular culture of items until an effort is made to remove that reference and replace it with a new one. The precise uses of a given word are peculiar to a language and its relationship with culture, leading us to the conclusion that language is a part of culture and that we can express cultural views and values through it.

In actuality, teaching a language always involves teaching a language and culture as well. Buttjest thinks that "knowing about culture is actually a critical component in being able to use and master a foreign linguistic system." "For effective international cooperation, knowledge of other countries and their cultures is as important as proficiency in their languages, and such knowledge is dependent on foreign language teaching," the Bellagio Declaration of the European Cultural Foundation and the International Council for Educational Development states.

Thus, acquiring a language entails knowing about the social norms and cultural practices of a particular society. A society's thought and actions produce its language. It is possible to think about teaching culture through learners' native languages, which can be used in a specific way to interpret the other culture (Taylor, 1979). A language speaker's effectiveness in a foreign language is directly correlated with his or her understanding of that language's culture (Ager).

Finally, we can draw the following conclusion regarding the speed at which cultural knowledge is acquired through immersion teaching: "...the integration of language and culture learning by using the language as medium for the continuing socialization of students is a process which is not intended to imitate and replicate the socialization of native-speaking teachers but rather to develop student's cultural competence from its existing stage, by changing it into intercultural competence" (Fengping Gao).

CHAPTER 6
Translation Differences Lead to Misunderstanding

Distinguished writer Arthur C. Clarke once noted that "Any sufficiently advanced technology is indistinguishable from magic." In my mind, we are beginning to witness this in our world of increasingly complicated projects.

The problem is that technical systems are getting increasingly pervasive, integrated, and arcane all at the same time. As a result, it is far easier for stakeholders to misunderstand issues and features. Misconceptions are very problematic because incorrect assumptions can result in difficult--if not impossible--project expectations.

Why "Magic"?

As systems and technology advance, assumptions will also start to be made by individuals more frequently. Experience, fad words, specialized jargon, and a host of other problems combine to make stakeholders confused.

Examples:

We will be able to access the software from any location because it is Web-based, it is assumed.

Project Team View: The program can only be viewed from within the firewall due to security issues. "Anywhere" is therefore a relative term that may not represent what the stakeholder has in mind in terms of a benefit or a problem.

It is assumed that since this new application is rules-based, anything can be made of it.

The project team's point of view is that we don't have the resources or the time to make it "do anything." In this case, stakeholders may have expectations that are outside the purview of the current project or phase of the project but could be satisfied given extra time and money.

Assumption: Since this application is the most sophisticated, it must also be the most user-friendly.

Project Team Perspective: Although it takes three times as long as your department's present application, it has significant advantages for the entire company. To assist the impacted group in planning for the changes, the project team will need to collaborate with the stakeholders.

These instances show how people frequently observe, hear, or read about something and then tend to draw conclusions based on rumors, previous encounters, what the system would behave if they developed it, etc. Additionally, if the product is advertised as having unlimited capabilities, laypeople may expect endless possibilities and develop an unachievable project scope.

The project team must take care to make sure that stakeholders are aware of the technical capabilities of any chosen solution and to properly record any challenges, dangers, etc. False assumptions have the potential to kill.

Comprehension and Communication

C&C, communication, and comprehension are at the core of this threat. The project team must actively seek needs, feedback, and comprehension levels in addition to communicating features. It is insufficient to merely publicize the capabilities of a prospective solution and solicit input.

It is necessary to assess and deal with the level of comprehension as well. The demand for precise understanding grows as project complexity and risk both rise. According to many, effective communication requires the two-way flow of information and assessment of communication effectiveness. This is accurate to some extent. The focus is that in addition to making sure that the key conversation topics are given effectively, extra effort must be made to guarantee that the arguments are grasped and no presumptions are created.

Terminology

To go further into the communication problem, it is possible to link the complicated world of technical terminology to a number of misunderstandings and subsequent failed projects.

Always keep in mind that talking to another engineer or technical party is one thing; communicating with them is quite another. It is acceptable to use technical phrases in that context, albeit perhaps with some clarification. Meeting a stakeholder from a non-technical area and then barraging them with technical jargon is ineffective, though. The person will either take offense, believe you are a stupid "techie," or believe your solution can accomplish something that it actually cannot. With regard to the latter scenario, depending on how far the mistake is allowed to persist and what decisions are based on it, it can be extremely difficult to rectify an incorrect assumption once it has been made.

Missing Context

People may comprehend specific language and have some level of comprehension, but they might not have the same amount of context that you do. This is another factor to take into account. In other words, they may comprehend the term's basic definition but not its significance to the project or your belief that they should hear your message.

What does "average cost" mean, for instance, if I want to value everything using that cost as a benchmark? Is it the mean of all expenses incurred since the initial purchase? Is it the mean of the 90 most recent days' worth of purchases? Is the average weighted? As you can see, there are almost endless ways to determine "average cost." Average cost needs to be understood by all project participants in order for them to use it effectively. Consider terminologies and acronyms that someone has never heard of; this is a simple term as well.

Embarrassment

The fact that people can be embarrassed to acknowledge they do not understand something just serves to exacerbate the problem of unclear wording. Nobody will inquire about an acronym or phrase that the entire audience may not grasp. Why? No one likes to appear foolish in front of their peers or, even worse, superiors.

Every form of communication involves a sender and a receiver. The message the sender wants to convey is expressed in terms that, in their opinion, most accurately capture what they are thinking. However, various factors can come into play and impede the intended message from being correctly understood.

Tone of voice can affect interpretation when communicating verbally. Hey, I saw you were taking an exceptionally lengthy break this morning," the supervisor might have observed, but if she or he had said it in a disapproving tone, the remark might have been taken as a slight reminder of the office's regulations. The remark may have even been a cordial question about what was going on and whether the employee needed any help if the person had a health issue that occasionally necessitates extended breaks. Here, the message's interpretation would be influenced by the speaker's tone of voice, the circumstances, and the relationships involved.

Additionally crucial are nonverbal indicators. Does the sender appear open and pleasant or aloof? Is she making a friendly or accusatory face? These variables all affect how the same words will be perceived.

Many other elements, in addition to how the communication is conveyed, affect how the message is received. Every new piece of knowledge we acquire is compared to what we already know. If it only confirms what we already know, we will probably appropriately interpret the new information, but we might not pay much attention to it. If it challenges our prior beliefs or understanding of the issue, we may twist it in our brains in order to match our worldview or we may just reject the information as false, misinformed, or simply incorrect.

When a communication is unclear, the recipient is especially inclined to clarify it for themselves in a way that meets their expectations. For instance, any unclear message will be interpreted as aggressive and hostile, even if it was not at all intended to be that way, if two individuals are engaged in an escalated argument and one assumes that the other will be hostile and aggressive. Our expectations act as lenses or blinders that skew what we perceive to conform to our preconceived notions of how the world works.

An example would be a study that looked at how people interpreted visual cues. People were forced to put up with upside-down images for a week or two after receiving eyewear that turned the world on its side. They were seeing things right-side up again once their brains learned to flip the visuals. When we hear something that we "know" is incorrect, the identical effect takes place. Our minds "correct" things so it looks the way we anticipate it to.

Misunderstandings are also more likely due to cultural differences. It is clear that improper translation poses a risk when there are multiple languages spoken. However, even when speaking the same language, people may still communicate differently.

Communication in high-context and low-context often differs from one another. Low-context communication can be understood on its own without the use of interpretation or context. The ambiguity of high-context communication is greater. For communication, prior knowledge and comprehension are just as important as the words themselves. Everyone communicates using both low-context and high-context methods, but Western societies tend to employ low-context methods more frequently than Eastern, Latin American, or African civilizations. If these distinctions are not acknowledged and taken into account, misunderstanding is almost certain.

Through influencing the recipients' assumptions, culture also has an impact on communication. As previously said, our minds attempt to distort incoming information to fit our viewpoint. Cross-cultural communication is especially prone to shift meaning between sender and receiver because various cultures have very distinct worldviews, and the sender may have a very different worldview from the receiver.

People in dispute sometimes misinterpret one other because we have a tendency to hear what we expect to hear. People will frequently want to partially conceal the truth because communication is already likely to be strained in these situations. As a result, there is a greater likelihood of misunderstanding and misconception, which can make managing or ending conflicts more challenging.

Tips for Preventing Misunderstanding

Avoiding miscommunication in conflict situations requires a lot of work. William Ury and Roger Fisher outline four abilities that can enhance communication in contentious circumstances.

Active listening is the first. They claim that the objective of active listening is to comprehend your adversary as well as yourself. Keep a close eye on what the opposing side has to say. Anything that is unclear or seems irrational (it might not be, but you are perceiving it that way) should be repeated or clarified by the opponent. Try to summarize their argument to them just as they have done so. This demonstrates your ability to understand what they have said while also demonstrating that you are paying attention to what they are saying. It neither implies nor requires that you concur with what they stated. All you have to do is say that you do comprehend them.

2. The second tenet of Fisher and Ury is to address your opponent directly. In certain

cultures, this is inappropriate, but where it is allowed, it promotes understanding. Avoid being distracted by other people or activities taking place in the same space. Concentrate on saying what you have to say in a way that your opponent can understand.

3. Talking about yourself instead of your opponent is the third of their three rules. Instead of concentrating on the intentions, wrongdoings, or shortcomings of your opponent, describe your own feelings and impressions. You can communicate the same message without making your opponent defensive or confrontational by using the phrase "I feel let down" as opposed to "You broke your word." Instead of utilizing "you-messages," this is frequently referred to as employing "I-statements" or "I-messages." You-messages imply responsibility and prod the recipient to downplay misbehavior or take the

blame themselves. I-messages don't assign blame; they only state an issue. This makes it simpler for the opposing side to contribute to the solution while avoiding having to acknowledge their error.

4. "Speak for a purpose" is the fourth tenet of Fisher and Ury. They caution that excessive communication can be detrimental. Consider what you want to communicate, why you want to tell it, and how you can do it in the clearest way possible before making an important remark.

To these four rules, more ones may be added. One is to use neutral language as much as you can. Inflammatory language rarely persuades listeners that the speaker is correct; instead, it tends to generate animosity and defensiveness. (In reality, it typically has the exact opposite effect.) Even though inflammatory statements might increase attention in a dispute and lead to support for a particular side, this support frequently comes at the expense of the conflict's overall escalation. It is best to make a point clearly without using strong language.

Similarly, respect should be shown to all opponents. Disrespecting others won't solve your problems; it will only make them enraged and less inclined to listen to you, comprehend you, or follow your instructions. If someone is treated with respect and decency, regardless of what you may think of them or whether you believe they deserve it, communication will be lot more successful and the problem will be easier to handle or settle. Deep talks (via problem-solving workshops or dialogues) can improve relationships, add context to communication, and dispel preconceptions that support unfavorable characterizations or worldviews, all of which can lessen misunderstanding. The likelihood that a communication will be appropriately comprehended increases with the amount of effort put into understanding the person sending it.

CHAPTER 7

Importance of Culture Awareness

Cultural sensitivity is to make sure that cultural variety may be effectively used to produce a successful project outcome. When project managers are sensitive to cultural differences, they recognize that others are unique and that, when used effectively, such differences may really be a source of strength. After all, everyone is a product of their society.

Through their: project managers assist in building connections that bring people together despite their differences.

1) Emotional IQ - Understanding how others are feeling

2) Capacity to Avoid Premature Judgment - Before making a decision, PMs must learn as much as they can about a problem.

With so many more individuals working from home and far fewer team members present on-site, it has become increasingly challenging to maintain open lines of communication over the past year. Nearly all meetings during this historic period were video chats or conference calls. Time zone differences frequently need to be taken into account on both a national and international level. Just getting everyone who is remotely located connected in real time might be a full-time task.

As challenging as it can be to keep everyone on the same page, managing a multicultural setting is considerably more difficult. PMs must constantly be aware of the different perspectives held by their team members due to their culture, way of life, and values. In a multicultural setting, things like gestures, eye contact, physical contact, language, nonverbal communication, and values might vary substantially. There are four phases of cultural awareness that people may go through, according to a recent article I read on "Business Analyst Learnings.com" titled The Importance & Stages of Cultural Awareness in Project Managers — Business Analyst Learnings:

• The Parochial Stage: the conviction that your method of doing things is the only valid one. You don't take cultural differences into consideration because you are unaware of your other options.

- The Ethnocentric Stage: You're aware that there are alternative methods to do a task, but you choose to ignore the cultural importance of a different approach because you believe your method is still superior.

- In the synergistic stage, you are aware of your options and decide on the best course of action based on the circumstances.

- The Participatory Third Culture Stage: You collaborate with other cultures to develop fresh approaches to addressing the demands of the circumstance.

Recognizing Diversity

Respecting those whose ideas differ from your own does not constitute cultural sensitivity. It involves more than just employing politically acceptable terminology. Understanding ethnic, socioeconomic, and cultural differences without passing judgment is the definition of cultural sensitivity. Different standards, communication methods, and behaviors are cultural considerations.

Project managers need to be aware of the participants' values and points of view when working in a group setting. Because culture affects individual viewpoints and mental processes, having this insight can result in effective teamwork.

When working with overseas clients, cultural awareness in project management might sometimes mean lowering communication barriers. Learning a few key phrases in the customer's native tongue or hiring a professional translator are two ways to improve communication.

A project manager must have excellent time management abilities in addition to a variety of other soft skills. The ability of the leader to embrace and value many cultures and ideas is the most crucial of these. When a team includes members from many cultures and values diversity, it can create an engaging work environment.

Consider a situation where a project manager is in charge of two teams from distinct nations. This implies that he or she should take into account the people's working cultures in addition to being aware of the various time zones and how to best promote communication between the two teams. While team members from the other country might jump right in and take on obstacles as they arise, people from one country may be used to working in a style that requires them to examine all the risks and comprehend the project before getting started.

There are many different methods, and one's method of working reveals a lot about their background.

Now that these teams are communicating, the manager can devise a strategy for getting them to cooperate while maintaining their distinctive working methods. The goal is to create a synthesis of the two cultures and working methods so that a consensus may be established.

A fusion model frequently raises the risk of cultural conflicts, which might prevent a team from reaching its full potential. The manager should be able to encourage the team members to communicate with one another and develop a strategy for minimizing these conflicts and promoting a positive work environment.

If we focus solely on the cultural component of this discussion, we can divide cultural traits into two categories: those that make up the more superficial aspects, such as language, dress code, eating habits, and so forth, and those that make up the deeper aspects, such as beliefs, sentiments, and values. Both play a crucial role in identifying a culture and a person within it. A manager can collaborate with everyone to develop a communication plan that benefits everyone.

Here are some instances of how to help culturally diverse work groups communicate more effectively:

Embrace a new point of view. Establish a setting where everyone is aware of how widely people's ideas can differ (about themselves and of others). Every person has a unique worldview. A manager can plan an activity that highlights the team's cultural diversity, such as organizing games that highlight the distinctive characteristics of other cultures.

Communication Techniques. If there are language difficulties at work, develop a communication strategy that everyone can understand and implement. For instance, if a team is predominately Chinese, ask one of the members to report on the meeting to the manager. Have the team have an internal status report meeting. No one will be unaware of what is happening in this manner.

Political Sensitivity

Professionals regularly deal with or take on the ongoing issue of organizational politics. Depending on a person's level of emotional intelligence, how they are seen might vary greatly. Project managers should be prepared to handle difficulties that may arise from team members succumbing to politics, which can quickly spiral out of control and impede the project's progress. Although organizational politics can be viewed in a variety of ways, the contrasts between those that hinder employees (such as hidden agendas or actions that injure a team member or a project) and those that can be seen as beneficial to the team and the project are the most obvious. The manager's main objective should be to perceive people's beliefs, to be attentive to their needs, thoughts, and opinions, and to use it to make good progress in uniting the team.

By carrying out duties and achieving goals in a practical manner, a transformational leader models leadership via example and practicality. His or her team should model behavior like this after them. The way a team should operate is determined by a transformational leader within constructive organizational politics.

In a politically charged workplace, there are several ways a manager might promote a positive working atmosphere:

Decide what caused it. Managers need to keep a careful eye on what goes on at work. This covers both typical events and any deviations that can impede the project's progress.

assist in communication Bring the team together to discuss the current issue and possible solutions. This will be to everyone's advantage. Give the team assurance that he or she is on their side and will protect them no matter what the issue. Any problem may always be resolved. Boost self-assurance. This reinforces my earlier point about building team confidence that, despite the stakes, their leader will be there for them, hold their hands, and lead them toward a successful project conclusion.

The defining pillars of an organization are its leaders, who also illuminate the way forward.

Years of managing teams and setting an example for others—not just by looking to other sources for those examples, but also from their own experiences—are necessary for developing leadership. A team or organization with a varied range of cultures is a terrific setting for learning, sharing, and exchanging important information and experiences. In the end, producing outcomes that can last a long time and show the project's success requires working in a setting that encourages everyone to speak up and offers them an equal chance to succeed.

Project managers need to pay close attention to multiculturalism in project management in the twenty-first century if they want to succeed. The issue that practically all project managers encounter on a daily basis is having a thorough grasp of other people who work with us from other nations. And because there are many levels of misunderstanding, this causes issues in management. This issue must be viewed as a challenge or a good thing to embrace rather than being run away from.

About 90% of the time that managers spend talking with their team members, stakeholders, or sponsors is spent on arranging and attending meetings, planning, identifying risks, negotiating, and resolving problems, to name a few. In general, project managers lack training in and sensitivity to cultural diversity.

Any group's culture can be described as a set of common characteristics that governs how that group arranges its interactions with one another, its surroundings, and how it responds to societal problems. The national character, perception, thinking, language, nonverbal communication, values, habits, and social grouping are just a few examples of how cultures differ from one another. Our society is so diverse now, which undoubtedly causes a lot of issues for the typical project manager. In most cases, the issue with verbal and nonverbal communication within a team is not only the result of Project Managers' (PMs) unwillingness to learn about the cultures of team members, but also the result of "cultural arrogance" on the part of individuals from other parts of the world who have settled in these nations and do not want to respect the values of their adopted country.

Minority women are present in our workplaces in large numbers and come from all over the world. These are some of the characteristics they exhibit at work:

- Standing up when a project manager approaches them with inquiries or problems.

- Looking down while a project manager is speaking to them rather than shyly is a cultural expression of respect!

- Avoiding interruptions when a PM is speaking.

- Maintaining a positive attitude even when things on the project are not going smoothly.

- They don't argue with you: This doesn't mean they agree with you or understand what you've said.

If project managers don't fully comprehend the aforementioned points, it frequently results in major issues.

The analysis of the four generations below adds another perspective to contemporary workplace issues:

- Veterans: It is generally accepted that persons who were born between 1922 and 1943 fall into this category.

Between 1944 and 1960, the baby boom generation was born.

People who were born between 1961 and 1980 were known as Gen Xers.

Born between 1981 and 2002, Nexters.

The first two groups, which may not be as diverse as the last two groups, include many of us in the USA. The majority of today's verbal and nonverbal communication-related project management issues arise from the first two categories' failure to fully comprehend the variations in team composition.

Gen Xers were born to working mothers and represent a variety of cultures. This group sees the world in a more unique way than the first two groups, and it typically understands multiculturalism concerns effectively and can settle disputes more rapidly.

The most productive of all the groupings may be the Nexters, who are quite cultural. Since they have been exposed to diversity their requirement for training in this area is little to nonexistent. They make up between 45% and 55% of the present labor force.

In order to make the workplace of tomorrow better than it is today, we must actually set an example for this final group of individuals: project managers.

To improve the future of Nexters, Gen Xers should be prepared and eager to collaborate with the remaining baby boomers and veterans in the workforce.

We all owe it to one another to give it our all today so that tomorrow will be better. Everyone will benefit from this in terms of higher workplace productivity, effective communication, meeting deadlines, lower costs, and a less stressful atmosphere.

CHAPTER 8

Persuading: Principles vs application

Robert Cialdini published his book "Influence: The Psychology of Persuasion" in 1984. In it, he explores factors that affect the decisions that people make, particularly in relation to sales and purchasing. His work is an influential precursor to Nudge Theory, and it's dark sibling, Sludge.

At the heart of his research is the now widely accepted notion that making decisions requires effort, which is why people frequently rely on heuristics and rules of thumb to decide what to do, how to act, or how to behave in various situations.

Particularly in respect to decisions regarding consuming and shopping, Cialdini has identified six fundamental principles that have an impact on these decision-making shortcuts. His key point is that if you comprehend these six ideas, you can utilize them to your advantage while trying to convince others to perform a certain action or purchase a certain item.

Cialdini identified six guiding principles: consensus, reciprocity, scarcity, authority, commitment, and consistency (or social proof).

1 - Mutuality

Reciprocity is the first of Cialdini's six principles of persuasion.

To some extent, people value fairness and balance (see, for instance, Adams' Equity Theory). This indicates that we dislike having a sense of owing others. In general, people endeavor to fulfill these social obligations when they have them. For instance, you almost surely want to give someone a birthday card in return if they send one to you. When their birthday comes around again, you'll do this to fulfill your social commitment.

This need for reciprocity can be used to sway other people's actions. To accomplish this, you must be the first to take action and provide the recipient with a unique and unexpected gift. The value of the present is somewhat less significant than the act of giving itself. The reason waiters give mints with the bill, workshop facilitators give cookies while they solicit input, and bosses give a team outing right before releasing the annual engagement survey is because of reciprocity. All of these gestures essentially communicate the message, "I've stroked your back, now scratch mine."

This reciprocity concept can be applied in the workplace by doing favors for others, offering assistance, complimenting others in public, and generally acting in a way that creates a bank of social obligations that are owed to you. At some point, all of these commitments will be resolved, possibly to your benefit. Of course, if you engage in this behavior excessively, it will stop working.

2 - Scarcity

Scarcity is the second of Cialdini's six principles of persuasion.

People tend to want things more when there are less of them available. This applies to both intangible experiences and tangible goods. Not much more needs to be said about this one.

This indicates that from the standpoint of persuasion and influence, you might profit by decreasing the availability of your good or service in order to raise interest in it (or at least creating a sense of scarcity).

This idea is put into practice in a variety of markets. For instance, online travel booking websites frequently state something like "just 5 seats left at this price." They act in this way to evoke a feeling of scarcity (as well as to add time pressure, which is closely related). Businesses in the consumer goods industry also create "limited edition" versions of products. They do this for goods ranging from shoes to hand soap, raising scarcity once more by reducing supply.

It might be able to generate a sense of scarcity around your own availability in the workplace. This can raise people's interest in what you have to offer. Of course, only people in positions of power are able to do this. When you don't have the authority, acting in that way may only result in people calling you ineffective.

Three: Authority

Authority is the third of Cialdini's six principles of persuasion.

Experts in their disciplines who are authoritative, reliable, and knowledgeable are more persuasive and influential than those who are not. This is partially due to the fact that trust is built on a number of fundamental tenets, including authority and credibility.

People are more likely to be obeyed when we have faith in them.

In many spheres of life, the authority concept is evident in action. White-coated dentists are employed to sell us toothpaste, airline employees wear uniforms to serve as a constant reminder of their authority, and many email signatures include a list of credentials in an effort to improve the sender's authority.

In truth, people who promote their own wisdom and authority are less successful than those who have others do it for them. Strangely, however, it almost makes no difference who that other person is. Your influence and persuasiveness are increased even if the individual endorsing you is known to gain something from doing so personally.

This means that while establishing trust and credibility is crucial in the workplace, it is also feasible to gain part of that sense of authority by receiving recommendations and glowing testimonials from others. It could be beneficial to ask others to recommend you or to recommend others so that they feel compelled to do the same for you.

4 - Consistency and commitment

People like to be consistent with their identity or sense of self-image, making commitment and consistency the fourth of Cialdini's six principles of persuasion. In other words, I'm more inclined to engage in behaviors that I perceive as being "healthy" if I consider myself to be a "healthy" person.

This means, from the standpoint of persuasion and influence, that if I can persuade you to take a small action in connection to anything, you'll think of yourself as that kind of person and be more likely to take that action again in the future. If I recommend that you boost your actions in that direction, you'll also be more likely to do so.

This can be compared to a salami-slicing persuasion strategy in certain ways. If I can persuade you to do one small thing, I can persuade you to do one smaller, similar thing as well. You'll perform a larger one after that. And before you know it, you've practically consumed the entire salami.

This kind of behavior is frequently seen in the marketplace with items like initial deals, which are simple and inexpensive yet serve as a doorway to other goods. Similar results can be attained with product giveaways. In the event that I offer you a free "World of Work Cookie" in the grocery store, you might begin to identify as a "World of Work Cookie Eater" and be more inclined to act in accordance with that identity going forward.

Commitment and Consistency in the Workplace - It might be able to use this principle to influence and persuade others in the workplace. To accomplish this, start by convincing others to do tiny things, then progress to bigger things. For instance, if I initially persuade my supervisor that "generally speaking some flexibility in working patterns is a desirable thing," she or he will be much more inclined to accept my proposed 4-day workweek when I later ask for it.

It's interesting to note that once someone does you a favor, they tend to identify as the kind of person that does favors for other people and are actually more likely to do so in the future. This means that if you can convince someone to do you a tiny favor at work (like lend you a pen or buy you coffee), they are more likely to do you another favor down the road. Of course, you have to live with yourself no matter how you choose to act toward and with other people.

5 - Enjoying

Liking is the fifth of Cialdini's six principles of persuasion.

Although it may seem intuitive, people are significantly more likely to be convinced and influenced by people they like than by people they dislike. Given the nature of humans, individuals who compliment and work with them are much more likely to be liked than those who don't. And, regrettably, despite strong data supporting some of the advantages of diversity, people are considerably more likely to favor those who resemble them than those who do not.

This approach is frequently applied in the marketing and advertising industries. The majority of the people in advertisements are made to appeal to the target market for the product. The consumer is more susceptible to being persuaded by someone they like and associate with more.

Simply become liked by those around you and those you are trying to influence or persuade in order to apply this approach in the workplace. You can achieve this through collaborating with others, giving others sincere comments, finding common ground, and forming connections. The important thing to remember is that before attempting to persuade others, you must first establish these connections and win their "liking." Once you've begun trying to influence someone, those efforts will fail if you try to win their favor.

6 - Agreement (social proof)

Consensus, or social proof, is the final of Cialdini's six principles of persuasion.

Humans are social creatures by nature, and most of us think it's necessary to follow the rules of a particular social group. This means that before making a decision, we frequently observe what others are doing in the immediate vicinity.

Towels from hotels are a well-known and well-known example of this. Signs that state, "Reusing your towel helps to save the environment," are considerably less persuasive than those that state, "8 out of 10 hotel customers opt to reuse their towels." It's interesting to note that the more socially focused these communications are, the more effective they are. Signs that, for instance, state that "8 out of 10 hotel guests who stay in this room opt to reuse their towels" are more persuasive than those that make no specific mention of hotel visitors at all.

From a personal standpoint, it can be challenging to apply the concept of social proof or consensus in the workplace, but it might be doable if you manage your reputation and personal brand.

Persuasion strategies frequently involve communication. The five canons of rhetoric and the rhetorical triangle provide some insight into how this operates. Explore Monroe's Motivated Sequence for a more thorough examination of persuasion communication. Additionally, the AIDA model is interesting to consider from a marketing angle. It could be interesting to read about trust and the Five Dimensions of Trust in Sales as a side study.

CHAPTER 9
Stakeholder Communication

What Is a Stakeholder in Project Management?

Stakeholders are those with an interest in your project's outcome. They are typically the members of a project team, project managers, executives, project sponsors, customers, and users. Stakeholders are people who will be affected by your project at any point in its life cycle, and their input can directly impact the outcome. It's essential to practice good stakeholder management and continuously communicate to collaborate on the project.

Stakeholders vs. key project stakeholders

Project stakeholders, in general, can be either single people or entire organizations that are impacted by how a project is carried out or how it turns out. Whether the project has a favorable or bad impact on them, if they are affected, they are a stakeholder. However, key project stakeholders are those who have the power and clout to decide whether a project is successful or not. These are the individuals and organizations whose goals must be met since they have the ability to make or destroy the project. The project cannot be deemed successful if the stakeholders are not satisfied, even if all deliverables are in and budgets are fulfilled.

Key project stakeholders that are typically:

The following are a few examples of the important project stakeholders you can encounter:

Customers: The people who really use a good or service, frequently both inside and outside the organization carrying out the project.

the person in charge of the project

Members of the project team: The team working on the project under the project manager's direction

The project's financial backer is the sponsor.

A steering committee is a decision-making advisory body that includes the sponsor, executives, and significant organization stakeholders.

Executives: The company's top management responsible for the project and for setting the organization's direction.

Resource managers are other managers in charge of the resources required to carry out the project.

There are numerous other instances of project stakeholders, including vendors/suppliers, workers, owners, administrators, and even the general public.

Megaprojects are typically very public in nature and have a broad-reaching impact on many diverse groups of people by their very nature. Due to the risk of losing support for the project and damaging the project owner's reputation, communication becomes even more crucial.

Through the course of the project's lifecycle, the following four communication techniques will help to ensure success:

Create a narrative that is consistent, identify all the stakeholders, choose the appropriate KPIs for external communication, and communicate risks proactively.

Identification of Stakeholders

A thorough awareness of your audience is the first step in effective communication. Start by compiling a list of stakeholders to accomplish this. Based on the strategic objective of your program, this tool will assist you in setting priorities for your time and effort. List every potential stakeholder, from the company's board of directors to individuals living in the areas where the project is being undertaken. Be thorough and detailed, and divide the list into groups.

Develop a distinct plan for communication and engagement centered on each stakeholder after carefully evaluating the unique needs of each group.

Build a Reliable Narrative

A narrative or story makes your thoughts and ideas more accessible to the audience. Before you need to explain the story, plan it out. It's also critical to understand that the plot may evolve over time. Until the megaproject is farther along, some components could need to act as temporary solutions.

A good story should address the following issues:

What is that? (State the project in plain, succinct terms, using examples from previous projects that are similar.)

• What problems will the initiative attempt to solve today? (This is the project's primary business case.)

• Why is this project crucial?

• What opportunities does the project present? (This should specifically address the difficulties.)

Why Why should the project be risked? Megaprojects have a history of cost overruns and delays, but these risks should be outweighed by their overall relevance.

How will the project be carried out? (This covers the participants and significant information about the project's goals and deadlines.)

- Which stakeholder groups will benefit most from the project? The crucial "what's in it for me" message is conveyed in this way.

- What measures will be taken to guarantee that the budget and timeline are met?

Decide on the appropriate metrics for external communication.

Determining how progress and success can be objectively monitored, as well as which metrics will be disclosed outside the project team and when, is one of the most crucial things to do in the early planning phase. The key three types of information that must be shared are:

Cost: Cost concerns are frequently at the heart of megaprojects. Frequently, when the level of accuracy is very low early in the design and estimating phase, project teams publicly release the estimated cost estimate. It is crucial to wait until the project is sufficiently specified before making cost predictions public, and then to communicate budget variance projected changes in cost estimations (across all stakeholders) as the design process progresses. Communication tactics should think about whether to use a range rather than a single figure to represent project estimates.

Schedule: A baseline schedule must first be created. The key to managing expectations and perceptions as scope modifications and unavoidable events (such as extreme weather, labor strikes, etc.) affect the project is routine, proactive communication across all stakeholders. Just as with costs, it's crucial to hold off on promising to achieve milestones or completion dates until the project's scope has been clearly defined, contracts with the contractors doing the work have been negotiated, and the risks associated with the project have been identified and investigated.

Manage Earned Value: The project team should update the stakeholders on project performance as it relates to the plan, forecast cost, and schedule performance.

Promote Risk Communication

Establish important benchmarks for evaluating and communicating risk, consult an expert as soon as possible, and don't forget to convey what you don't know. Operating in a culture that prioritizes risk management is also beneficial since hazards cannot be successfully detected and conveyed without active risk management.

Project owners can lessen the potential harm that a lack of awareness about the project could create by recognizing communication as a basic part of designing the project and adopting effective communication tactics.

Approaches to Handling Difficult Stakeholders

It's far simpler to say than to do to design a strategy for project stakeholder management. Stakeholder management can be difficult, especially if they act in ways that compromise the common goal. We've compiled some telltale signals that you might be working with a troublesome stakeholder to assist you in identifying them:

communication problems

Taking note of stakeholders' expectations and project objectives is one of the most important components of managing stakeholders. You will need to specify performance standards, project limitations, and additional information regarding how stakeholders define project success at this stage of the project. This is how you may assign assignments with more assurance.

Regardless of your field of business, efficient communication is essential to any firm. In light of this, stakeholders who don't promptly answer phone calls or respond to emails don't help make your job any easier. This silence might also be seen as apathy in the endeavor as a whole.

They simply offer criticism.

Do your stakeholders always provide negative feedback when they express an opinion? Another red flag that you're working with challenging stakeholders is this: During stakeholder meetings for your project, you're more likely to hear criticism.

Knowing what to say to a client who is dissatisfied with the project status report can be very challenging. Yes, constructive criticism is always beneficial, and occasionally, you could even gain from being boldly honest. However, some participants might be unpleasant and unhelpful without good reason.

They don't feel the same urgency.

From the beginning to the end, everyone of your stakeholders should feel the same level of urgency. After all, you'll have more money in your pocket if you finish sooner! There may no longer be a sense of urgency in the minds of your stakeholders if they appear to be putting the brakes on your project or providing input slowly. A big red signal is when stakeholders fail to convey a feeling of urgency.

They prematurely withdraw resources from the project.

Functional managers and other stakeholders may not necessarily perceive the benefit in the project if they refuse to release the resources needed to launch it.

As a result, there may be delays and poor quality with your project. Your best option in this situation is to strive for total campaign transparency, which can help increase stakeholder confidence in the initiative.

Remember not to cross any bridges.

The most crucial thing to keep in mind is that stakeholders are interested in the project's success as well. However, as the project progresses, individuals might alter how they convey this goal. If work isn't done a certain way, they'll encourage you one day then criticize you the next. The success of the initiative is their side, thus they are not "moving sides." You and them are not at war. Do not take opposition personally and always keep in mind that business is business. Building fewer relationships is harmful for career success. You cannot ignore challenging stakeholders. To diffuse the issue, you must find a method to cooperate with them (or get around them).

1. Keep an eye on them.

Finding out who your stakeholders are and what drives them is the first step. Anyone who is impacted by your work, has control over it or an interest in its success is a stakeholder.

Stakeholders fall into three broad categories:

principal players

People who are directly impacted by the work are the main stakeholders. Typically, they are project beneficiaries. Clients frequently fit into this category.

second-tier participants

People who indirectly benefit from the work are considered secondary stakeholders. Teams supporting the project and those who will be touched by its results are considered secondary stakeholders.

Key parties involved

Key stakeholders are those who have a significant impact on the work and a stake in its success. Executives are included in this group. Each organization has unique interests, goals, and agendas, many of which are at odds with one another. To keep initiatives moving and prevent being pulled in a number of different directions, identify and rank their influence and interest. Finding the influential person who will be your best ally will relieve a lot of your worry because not all stakeholders are created equally.

a) Manage highly influential and motivated individuals carefully.

These individuals are very interested in your job and have the ability to make you successful. It's crucial to involve them and make sure they're happy. Before beginning a new project, talk to them, listen to what they have to say, and, wherever you can, put their suggestions into practice. When someone else's suggestions are chosen, let them know and explain why.

b) High-power, uninterested individuals: maintain satisfaction

These folks are incredibly influential while having minimal engagement or personal investment in your job. Try your best to keep them happy, but don't monopolize their time. Ask for their opinions when making important decisions and make sure they are aware of the beneficial effects your effort will have on them. Once you get their support, these people become formidable allies.

b) Low-power, intensely curious individuals: stay informed

Despite having little influence or power, these individuals are fervent about the cause and publicly express their support. Keep them updated and let them know about any significant developments. These folks may be immediately impacted by your job, therefore they are frequently more than happy to pitch in and assist you.

c) Low-power, uninterested individuals — observe

These individuals should not require much of your time or attention because they are the least affected by your job and are the most apathetic of the group. If you don't upset them, they'll remain out of your way.

2. Take in what they have to say

If you don't like what you hear, don't cut off communication. Put yourself in the position of challenging stakeholders to better comprehend their motivations and objectives. Try to understand where they are coming from.

Make an effort to comprehend their viewpoint. If what they're saying frustrates you, consider whether their requirements match the goals of your project. Do they merely prefer a different method of doing things? Look for points of agreement.

People desire understanding and appreciation of their thoughts above anything else. Here are some strategies for managing stakeholders and demonstrating their importance:

Always treat individuals with respect, even when tempers flare. Praise frequently, especially when you observe excellent behavior. Provide training and coaching to all parties involved. Give people opportunity to share their ideas and opinions with the group and assist in decision-making.

3. Go see them one to one

Plan a meeting with each of the challenging stakeholders. They feel less pressure and more at ease while meeting alone, away from other stakeholders. Conversations become more lucid and relaxed as a result.

Explore their point of view and suggested solutions at this time. Don't, however, question them outright why they disagree with your strategy. Ask them open-ended questions instead to learn more about their viewpoints and how they believe the project is coming along.

Managing stakeholders one-on-one also prevents their unfavorable viewpoints from having an impact on other project participants. The best course of action is to isolate the stakeholder and handle the problem privately when input crosses the line from constructive to purely negative.

4. Establish what drives them.

Why are your stakeholders suddenly becoming resistant? Concerned about exceeding the budget? Concerned that the project isn't going as planned? Do they answer to a board of directors that has its own reservations?

The secret to managing stakeholders is to deal with the reasons behind their resistance; doing so will enable you to identify compromises, come up with a win-win solution, and complete the project.

Successful stakeholder management involves many different elements, one of which is picturing project timelines and objectives. Project managers can gain more visibility into the output and workload pressure of their teams by using Wrike's sophisticated reporting tools.

You may get rid of communication bottlenecks and make sure that your team and stakeholders can work together to accomplish business goals by using automation, real-time data, and configurable dashboards. So that you and the project's stakeholders are always in sync, centralize project management onto a single platform.

CHAPTER 10

Trusting: Task based vs Relationship based

Trusting scale has two dimensions – Task-based and Relationship-based. As you can probably guess, the task-based dimension is more about the 'you do good work consistently, you are reliable, so I trust you'. The second dimension on the other hand is rather the approach of 'I've shared my personal time with you, I like you, I know others who trust you, therefore I trust you'. As much as you can like someone who does the work well and consistently, the trust is not based on your 'liking' in case of a Task-based preference. It works the other way around as well – the alone fact, that you do a good work may not be enough for me to trust you if I'm from a Relationship-based culture.

In order for multicultural teams to be effective, they need to establish trust, and how we build trust can differ greatly from country to country. Understanding your own preference and that of others when it comes to team building can mean the difference between success or failure as a team.

Erin Meyer describes the dimension of trust as cognitive trust vs. affective trust. Cognitive trust is based on the confidence you feel in another person's accomplishments, skills, and reliability. This is trust from the head. Affective trust arises from feelings of emotional closeness, empathy or friendship. This is trust from the heart.

In addition to understanding if we trust from the head or the heart, we also need to understand task vs. relationship-based trust. Someone with a task-based approach prefers to build trust by understanding the task and getting to it. Someone who prefers a relationship-based approach needs to understand the people with whom they are doing the task.

For task-based cultures, trust is built through business-related activities. Work relationships are built and dropped easily based on the practical needs of the situation, and trust grows when deadlines are met, and goals are achieved. In relationship-based cultures, there is greater emphasis on building trust through sharing meals, engaging in meaningful conversations, sharing personal interests and concerns, etc. Work relationships build slowly over the long term. Someone with a relationship-based approach to trust building might think, "I've shared personal time with you, and I know others who trust you. Therefore, I trust you."

Cognitive Trust & Affective Trust

Before we dive into the trust scale, it's worth quickly touching on the difference between cognitive & affective trust, as these impact our understanding of the trust scale:

Cognitive Trust:

- Trust is based on the confidence you feel in another person's accomplishments, skills & reliability.
- Trust comes from the head.

Affective Trust:

- Comes from feelings of emotional closeness, empathy or friendship.
- Trust comes from the heart.

The Trust Scale: Task-based vs. Relationship-based Trust

Now let's look at the high-level differences between the two ends of the trust spectrum:

Task-based Trust:

- Trust is built through business-related activities.
- Work relationships are built and dropped easily, based on the practicality of the situation.
- General mentality = "You do good work consistently, you are reliable, I enjoy working with you ... I trust you."

Relationship-based Trust:

- Trust is built through sharing meals, evening drinks & visits at the coffee/tea station.

- Work relationships build slowly over the long term.
- General mentality = "I've seen who you are at a deep level, I've shared personal time with you, I know others well who trust you ... I trust you."

The Relationship between Cognitive/Affective Trust & Task-based/Relationship-based Trust

Generally speaking, ...

- The further a culture falls toward the task-based end of the scale, the more people from that culture tend to separate affective & cognitive trust ("business is business" ...leave personal life out of it).
- The further a culture falls toward the relationship-based end of the scale; the more cognitive and affective trust are woven together in

business ("business is personal... i.e., you can't do business without a relationship).

How Confusion Arises & How to Deal with It

It can seem strange to "Coconuts" that "Peaches" would so willingly share personal information about themselves and ask what are very personal questions in a Coconut culture.

- If you're a peach, be aware of whether or not you're talking to a peach or a coconut, and how your everyday questions might be perceived as personal. If working with a coconut, spend some time getting to know them in a non-work setting and you'll find that Coconuts are more than happy to open up once they know you on a personal level.

It can also seem very disingenuous when a Peach has such a personal conversation with a Coconut and then they leave the conversation without any sort of indication that they intend to see or talk to the Coconut again.

- This is very normal for Peaches, it's nothing personal. If a Peach "overshares" with you, it's not their way of asking for a deep relationship and they're not trying to mask some sort of hidden agenda. It's just a cultural norm for a Peach to act in this way.

Finally, it can be very bizarre for a Coconut to have a Peach be so open about so many things and then run into their "pit" and be blocked off from actually getting to know who the Peach is at a deeper/more real level.

- Again, fellow Peaches, remember that Coconuts are relationship-based. If you are willing to share big parts about your life (about your family, what you did on the weekend, etc), don't be afraid to share more personal information.

Coconuts → Peaches

It can be offputting to Peaches when they attempt to be friendly with a Coconut and receive a "flat" response in return (or something that is not perceived as being equally friendly).

- Coconuts, your Peach co-workers aren't weirdos. We are just trying to get to know you in our own way.

- Peaches, your Coconut co-workers aren't snobbish, arrogant, or angry with you. This is just weird for them.

Peaches: Be yourself; smile as much as you want, share as much as you want, but don't expect Coconuts to do the same & don't ask personal questions until you've gotten to know the Coconut on a more personal level (& remember, this takes time and non-work interactions and activities).

Strategies for Success

Working with Task-Based Cultures:

- Don't throw out the socializing altogether. Task-based peers also enjoy socializing, meals, outings. However, if it is going to be a "long lunch" (over an hour long), tell them about it in advance.

- If you invite a task-based colleague out for drinks after work, and they decline so that they can go back to the hotel and rest or catch up on emails, don't take offense .. this is a normal and an acceptable response in task-based cultures.
- For communication: Choose the method that is the most efficient - email, chat, video conferencing, etc are all acceptable as long as the message is communicated clearly and succinctly.

Working with Relationship-Based Cultures:

- Be willing to share your personal/non-work self (it's OK to "let loose" and have a few drinks, or go out for karaoke and sing your little heart out). You don't always have to be "professional." This can be misconstrued as uncaring, not having fun, disrespectful, etc

- Act as if you're out on the town with your best friend.
- Don't worry about saying or doing the wrong thing. Just be yourself - your personal self, not your business-self.
- Put more time & thought into organizing meals or other events to create moments/memories together.
- Don't think of "long lunches" or long coffee breaks as a "waste of time" → You're investing in your relationship which will pay dividends for how you work with the person/people for a long time.
- For communication: Choose a communication medium that is as relationship-based as possible (instead of email → chat; instead of chat → video conference; etc.)

Working on a Global Team with a Mix Cultures:

- Invest in building up your affective trust → it's worth the investment.
- Build on common interests:
- "You like music? Me too...what are some artists you enjoy?"
- "Yeah, I have kids too. It can be challenging, but it's amazing."
- If it's hard for you to find common interests... look harder for them; they're there :)

CHAPTER 11

Leading Across Cultures (Hierarchical vs. Egalitarian Leadership)

Cultural differences in leadership styles often cause unexpected misunderstandings.

Americans, for example, tend to consider themselves egalitarian and think of the Japanese as hierarchical. But American leadership can be confusing. Though American bosses are outwardly egalitarian—asking subordinates to use first names and to speak up in meetings—they can be extremely top-down in the way they make decisions.

How do leadership styles differ across cultures? Why is it important to understand a culture's approach to leadership when working internationally?

According to cultural communication expert Erin Meyer, leadership styles differ drastically across cultures. If you're a manager, Meyer argues that it's your responsibility to adapt your leadership style to what your subordinates are used to. But even if you're not, knowing how leadership works in different cultures can help ensure you give a good impression and don't offend anyone.

We'll look at how Meyer divides cultural leadership styles into two extremes (egalitarian and hierarchical) and present some strategies for working and leading in different kinds of cultures.

Power Distance & Leading

In the 1970s Geert Hofstede coined the term "power distance", which refers to *the extent to which the less powerful members of organizations accept and expect that power is distributed unequally.* The Leading Scale takes the idea of power distance and applies it specifically to business, with the two ends of the spectrum being egalitarian leadership & hierarchical leadership, respectively.

Power distance relates to questions like the following:

- How much respect or deference is shown to an authority figure (and their opinion)?
- Is it acceptable to skip layers at your org?
  → i.e. If you want to communicate to someone two levels above or below you, should you go through the hierarchical chain?
- When you are the boss, what gives you an aura of authority?

Egalitarian vs. Hierarchical Cultures

In her book The Culture Map, Erin Meyer defines two cultural leadership styles: egalitarian and hierarchical.

The leadership style preferred in a country reflects the amount of "power distance" expected in that country. Power distance, a concept introduced by pioneering cultural theorist Geert Hofstede, measures how hierarchical a country is and how its citizens value authority. The citizenry's expectations of how power is distributed determine whether a country prefers an egalitarian or hierarchical leadership style.

In an egalitarian culture, the power distance is low. In other words, everybody is equal—even in the workplace. Companies in egalitarian cultures tend to have a flat organizational structure. People speak as easily to the CEO as they do to the lowest-ranking employee.

In a hierarchical culture, the power distance is high. In other words, your rank matters. Companies in hierarchical cultures have clearly defined levels, and the employees stick to them. They talk to their immediate boss and subordinates but receive permission to talk with anybody further up or down the chain.

Hierarchical Leadership (High Power-distance):

- The ideal distance between a boss & subordinate is high
- The best boss is a strong director who leads from the front
- Status is important
- Organizational structures are multilayered and fixed
- Communication follows set hierarchical lines
- Leaders display their leadership/authority by setting themselves apart from those at lower levels of the organizational ladder

Egalitarian Leadership (Low Power-distance):

- The ideal distance between a boss & subordinate is low
- The best boss is a facilitator among equals
- Organizational structures are flat
- Communication often skips hierarchical lines.
- Leadership looks more like "acting like one of the team" (and not separating yourself from your subordinates)

General Strategies for Working in Hierarchical Cultures

If you need to speak with someone several levels above or below you, always get permission from the person in between. Then, make sure to copy that person on any emails you send. This way, your email recipient understands that they can respond freely without violating local business etiquette.

(Shortform note: Pay attention to your tone in emails—especially when emailing someone of similar rank. Don't write anything that might be interpreted as bossing them around. People in hierarchical cultures are particularly sensitive to suggestions of unearned authority.)

In hierarchical cultures, there are often subtle ways in which rank is expressed in business. For example, in Japan, the lowest-ranking person always operates the elevator buttons, while the highest-ranking person stands directly behind him/her. Learn these rules well so you don't unintentionally offend your colleagues or clients. You should also learn the non-subtle ways too, like proper bowing etiquette in Asian cultures.

Err on the side of caution and refer to people by their last name unless they indicate otherwise. And don't insist that people call you by your first name, since this can introduce unnecessary discomfort in your relationship. Consider a compromise like "Ms. Jane" instead. (Shortform note: Many hierarchical countries, like South Korea and Japan, have language tenses that indicate politeness—you speak, quite literally, differently to someone based on whether they're above or below you in rank. If you're speaking English, referring to people by their last name or having them refer to you by their last name may be one of the few immediate ways you can show respect.)

How to Lead in a Hierarchical Culture

Take your responsibility to protect seriously. Your subordinates' faithful obedience doesn't give you license to treat them poorly. Hierarchical leadership works best when the leader protects and mentors their subordinates well. (Shortform note: Of course, the idea that leaders need to protect their subordinates isn't exclusive to hierarchical cultures, with one article attributing it to the American 1967 business book Organizations in Action.)

When you need your team's input, tell them before the meeting happens. If they know in advance that you want their honest opinion, they'll do their best to respect you by providing it. They'll also have more latitude to consult with their colleagues. (Shortform note: You could also consider soliciting anonymous opinions. Try setting up a dummy email account or Google document where people can express their thoughts privately.)

Even if you've requested people's input beforehand, call on people whose opinions you want to hear in meetings. People used to hierarchical leadership styles tend not to volunteer their input unless specifically asked.

(Shortform note: But skip this strategy if you're an external presenter. Calling on people in front of their bosses could embarrass them.)

Alternatively, consider removing yourself from the meeting entirely. Instead, have someone present the meeting's conclusions to you later on. Your subordinates will feel more comfortable expressing their ideas honestly if you're not present. (Shortform note: But remember that there may still be hierarchical relationships at play that affect the meeting results. Hierarchical relationships can be based not just on job title but also on factors like gender, age, or years at the company.)

General Strategies for Working in Egalitarian Cultures

Remember that speaking with someone several levels above or below you is likely totally normal—no matter how uncomfortable it makes you feel. (Shortform note: If you're particularly intimidated by somebody, try comic visualization: Picturing them in a funny situation encourages your brain to turn its stress response off.)

Copy people on your emails on a need-to-know basis. Copying the boss unnecessarily may make an egalitarian employee feel like you're trying to make them look bad or that you think they need extra oversight. (Shortform note: Some people worry that if they don't copy their boss on emails, they can't keep their boss informed. Try some other strategies, like sending your boss occasional recaps of your exchanges.)

How to Lead in an Egalitarian Culture

Learn and adopt the external cues that indicate that you're 'one of the guys.' For example, call people by their first name and insist they call you by yours. Using last names may be overly formal and stiff to someone from an egalitarian culture. (According to Meyer, this is true of Scandinavia, the Netherlands, and Australia, but can vary regionally in the United States and the United Kingdom.) (Shortform note: This may reflect the psychological concept of social mirroring, the idea that we unconsciously copy people we like. Many texts recommend mirroring others' gestures consciously in an attempt to get them to like you.)

Meyer also recommends several strategies that fall under the 5-step method of Management by Objective, a framework developed by management expert Peter Drucker in the 1950s. Although she mentions the term "management by objective," Meyer doesn't present her strategies using the step-by-step sequence Drucker did. We, however, will use the 5-step framework because it's easier to understand.

Step 1: Set goals for your team.

Step 2: Share these goals with your team.

Step 3: Ask your employees to set their own goals. Approve them as long as they support the team objectives you've shared. Meyer recommends this specifically because people from egalitarian cultures feel more comfortable with a facilitator rather than the traditional hierarchical leader. Similarly, she recommends facilitating instead of leading meetings, too.

Step 4: Monitor your employees progress on those goals.

Step 5: Assess your employees progress and reward them as necessary.

General Traits of these Cultures:

Hierarchical Cultures

- Disagreements with the boss: An effort is made to defer to the boss's opinion, especially in public
- Need for boss's approval: People are more likely to get the boss's approval before moving to action
- Need to match hierarchical levels: If you send your boss, they will send their boss. If your boss cancels, their boss may also cancel
- "Level Skipping": Communication must follow the hierarchical chain

Egalitarian Cultures:

- Disagreements with the boss: It's okay to disagree with the boss openly, even in front of others
- Need for boss's approval: People are more likely to move to action without getting the boss's approval

- Need to match hierarchical levels: If meeting with a client or supplier, there is less focus on matching hierarchical levels
- "Level Skipping": It's okay to email or call people several levels above/below you

Avoiding Confusion

Hierarchical → Egalitarian Teams

- Confusion or hard feelings can occur if you

don't include your team members in decision-making. If you're working with egalitarian team members, seek their input (you can do this while still holding the final say, see the 'strategies for success' section below).

- If team members disagree, it's not a sign of disrespect - this is part of how egalitarian cultures come to consensus.
- If a hierarchical team member gives you a plan without asking for your opinion/ideas, it's not a sign of disrespect → This is just more common for their culture.

In hierarchical cultures, the leader's responsibility for caring for and teaching his subordinates is just as strong as the follower's responsibility to defer and follow instructions. It's a system based on reciprocated obligations, not just the obligations of "the underlings".

Egalitarian → Hierarchical Teams

- Confusion or hard feelings can occur if you are "skipping levels" in the org. chart. If you're a manager, discuss issues or ideas with your peers. At a minimum, ask your peer if it's OK to discuss issues/ideas with their team members.

- If an egalitarian skips hierarchical levels, it's not a sign of disrespect → This is just common for their culture.

Strategies for Success

Working with Hierarchical Cultures

Strategies for Working with Hierarchical Cultures

- Communicating with 'the source' or the boss: Communicate with the person at your level. If you are the boss, go through the boss of the equivalent status, or get explicit permission to hop from one level to another (otherwise, you run the risk of offending your peer).
- Including bosses in emails: If you email someone at a lower hierarchical level than your own, copy their boss.

- "Level Skipping": If you need to approach your boss's boss or your subordinate's subordinate, get permission from the person at the level in between first.

Strategies for Working with Egalitarian Cultures

- Communicating with 'the source' or the boss: Go directly to the source; No need to bother the boss.
- Including bosses in emails: Think twice before copying the boss on emails. Doing so could suggest to the recipient that you don't trust them or are trying to get them in trouble.
- "Level Skipping": Skipping hierarchical levels is generally fine.

Tips for managing teams in different cultures

Managing Hierarchical Cultures

If you are managing a group that defers to your authority/opinion but you want them to provide input so you can make informed decisions, try the following:

- Ask the team to meet without you to brainstorm as a group & then bring ideas back to you (removing "the boss" from their team meeting allows them to feel more comfortable sharing ideas & then they can share these ideas as coming from "the group" rather than any one individual).
- When you have a meeting coming up, give clear instructions to the team in advance as to what kinds of questions/issues will be discussed and let them know that you will be asking for their input (this allows them to organize thoughts & check-in with one another prior to the meeting → see above)

- Don't expect people to jump in randomly without invitation; instead invite people to speak up and share ideas → Even if team members have ideas prepared, they might not volunteer unless you call on them to share.

Managing Egalitarian Cultures

If you are managing a group that is more egalitarian in nature than you are used to, try the following:

- Introduce management by objectives, starting by speaking with each employee about the team's vision for the coming year and then asking them to propose their best personal annual objectives, subject to negotiation and final agreement with you. This allows you to become a facilitator rather than a supervisor while still keeping a handle on what is being accomplished. *Make sure the objectives are concrete & specific and consider how to link them to rewards/recognition.*

- Check on progress periodically (for a year-long project, maybe once a month so as to not be seen as micro-managing). If progress is satisfactory, you can give your team member more space for self-management; if progress is lagging, you can get more involved. *It always helps to set goals & expectations for how this will work up-front too, so don't hesitate to do that.*

CHAPTER 12

Providing Feedbacks Across Cultures

Working on a team that is globally distributed is a great experience which also presents opportunities for deeper analysis and understanding. Since SPS is a multi-national, and thus multi-cultural, company, I will be writing eight individual blog posts which follow the eight scales of cultures outlined in Erin Meyer's book, The Culture Map.

The Evaluating Scale: How Negative Feedback is Provided

- Direct Negative Feedback: Negative feedback is *frank, blunt, & honest*. Negative messages standalone (not softened with positive messages). Absolute descriptors are common (*"always", "never", "absolutely", etc*). Criticism may be given to an individual in front of a group.
- Indirect Negative Feedback: Negative feedback is *soft, subtle, & diplomatic*. Positive messages are used to "wrap" (or soften) negative ones. Qualifying descriptors are common (*"sort of", "maybe", "slightly", etc*). Criticism is given only in private.

How Others Might Perceive You...How Misperceptions Arise

Direct Feedback Cultures

- Others can have a perception that Ukrainians (and other direct feedback

cultures) are rude, mean, harsh ("Why does this person only ever seem to point out the mistakes I make and never recognize the contributions I provide?")
- There can also be a perception that something is a personal attack if negative feedback is given with others present (it's also really uncomfortable for your indirect feedback colleagues when this happens)

Indirect Feedback Cultures
- Others can have a perception that Americans (and other indirect feedback cultures) are false/dishonest or just plain confusing ("Why are you telling me so many things that are great, if you really want me to focus on something that I need to fix?").

Being aware of these perceptions can help us to know both: 1) how we are perceived by others, and 2) how we are perceiving others so that we can adjust accordingly. This will often come through open dialogue with your team members so that you can better understand where the other is coming from when providing (or receiving) negative feedback. Without open dialogue, misperceptions can turn into held beliefs, which can turn into resentment of others. Don't let this happen!

Strategies for Working with Indirect Negative Feedback Cultures

Giving Feedback TO Indirect Feedback Colleagues:
- When providing an evaluation, be explicit and low-context with both positive & negative feedback → BUT, don't jump straight to the negatives. Make sure you

state something positive about the person/situation first. Note, these positive comments should still be honest & explicit.

- Try replacing "upgraders"/absolute statements with some qualifiers or "downgraders" (words to soften the feedback). For example, instead of saying something like "You always …" or "You never…", try saying something like "It seems like you occasionally…" or "I've noticed that sometimes…" ← (I know, it's weird).
- Work to eventually become balanced in the amount of positive and negative feedback you give.
- Only give negative feedback in private (not in front of others) → This includes private chats vs. Slack rooms. For less individualistic cultures (i.e. collectivistic

cultures), like India, you should also apply this rule to positive feedback as it may make these co-workers uncomfortable being called out in any way in a public setting. Your American colleagues, however, love being praised in public, so keep it coming!
- If your colleague is also a high context individual, then try blurring the message, giving it slowly over time, and sharing over drinks/food (when possible)

Receiving Feedback FROM Indirect Feedback Colleagues:
- Remember that your colleagues aren't intending to be false or confusing. We Americans have grown up being taught that the best way to give negative feedback is to couch it with loads of positive feedback and qualifying

statements. That isn't true for people who have grown up in a more direct negative feedback culture.

Strategies for Working with Direct Negative Feedback Cultures

Giving Feedback TO Direct Feedback Colleagues:

- Don't try to "do it like they do" (in terms of "upgraders", especially) → It's possible to be too direct, and if you're not from that culture, you won't likely know what that line is. If anything, working in one "upgrader" may be safe ("You're completely wrong. ", "That's a terrible idea.", etc). If unsure, don't try to include upgraders. Keep it simple and just be more direct.

- Don't use your fallback Compliment Sandwich method (or 3 positives per negative) → This is confusing to your direct feedback colleagues, as it actually sounds like you're telling the person they're (doing) amazing!
- Try limiting your statements that could be seen as superfluous or exaggerations: For example, you don't have to use so many "great", "excellent", "amazing", "thrilled" type words → If you do, do so sparingly. Otherwise, you will again run the risk of being confusing or being seen as false with the feedback you're giving.

Receiving Feedback FROM Direct Feedback Colleagues:
- Take messages literally and not personally → They're not intended to be offensive; take the message at face value.

- Understand that direct negative feedback is a sign of honesty, transparency, and respect for your professionalism (i.e., you can handle it and you should want to hear it so that you can improve).

No matter who you are interacting with and what their cultural norm is...If it ever seems like something is unclear, ask clarifying questions until you are both on the same page. I.e., don't assume that your message was received as it was intended to be. It's better to seek clarity and know that you're on the same page than to avoid doing so and fall into miscommunications, misunderstandings, and poor feelings towards, or about, your colleague. Negative (constructive) feedback is a valuable part of helping one another to grow and develop. Hopefully using these tips, you will feel more equipped to manage giving and receiving negative feedback with your cross-cultural peers.

CHAPTER 13

Benefits of Global Diversification

The benefits to the business are relatively straightforward: There are generally cost benefits of having teams distributed across the globe. If done right, it can also provide 24/7 coverage and coverage during major holidays. When it doesn't require late night and weekend on-call rotations, being able to continually ship code to production is a significant win for any tech team, whether that's new deployments or bug fixes.

The advantages don't stop there. Global diversity (and diversity in general) can benefit your team in a variety of ways, including the following: People from various backgrounds naturally approach challenges from various viewpoints and techniques. People from various backgrounds also frequently learned to code using a variety of programming languages, best practices, etc. At first glance, this can seem like a "negative" thing. After all, wouldn't life be better if everyone had the same perspective and behavior? As it turns out, and this shouldn't come as a surprise, that isn't the case. The more diverse the experience and training resources, however, the more resilient your team will be when dealing with the constant barrage of unique problems that arise when working as a tech team. Even if the topic your team is working on is not urgent, having a diverse group of programmers and problem-solvers can help them generate more (and more original) solutions to the numerous issues they are always trying to address.

After a significant rise in the U.S. stock market since early May, market volatility spiked back in October. On October 11, the S&P 500® Index of large company stocks in the United States fell over 7% from its peak as a result of worries about trade and increasing interest rates. The resurgence of volatility was a powerful reminder that financial markets are inherently erratic and that various asset classes experience ebbs and flows of popularity throughout time. Because of the potential for abrupt changes in the market, diversifying one's portfolio internationally is crucial. Consider defensive asset groups like gold, which went from being a follower to a leader and delivered gains as markets declined. While short-term volatility might be upsetting, it's crucial to keep in mind that one of the keys to long-term investment success is maintaining diversified and focused on your longer-term goals as different asset classes move up and down the performance rankings.

Remember that when analysts discuss whether "the market" is up or down, they often refer to indexes like the Dow Jones Industrial Average or the S&P 500® Index and their performance. Investors should be aware that these frequently referenced indices only represent a portion of the entire market. The Dow contains only 30 businesses, whereas the S&P contains around 500. Both are measures of large-cap U.S. stocks. In reality, until the end of September, the main U.S. large cap stock ETF used in Schwab Intelligent Portfolios® had advanced by nearly 10.6%, keeping pace with the S&P 500 Index.

However, the global stock market also comprises thousands of companies from U.S. small-cap stocks, international developed markets, and developing economies in addition to U.S. large-cap stocks. In addition to stocks, the market also features different kinds of bonds, commodities, real estate, cash, and other investments. Investors need to understand how each of the indexes that measures these distinct market categories has performed in order to comprehend how the entire market has performed and how it can affect your diversified portfolio.

Benefits from diversification include:
Beyond equities, wise investors make other investments. The S&P 500's "correction" of more than 10% in February 2018 and a fall of more than 55% during the financial crisis show how unpredictable U.S. large cap stocks can be. While equities fell during the financial crisis, Treasury bonds led the market in performance with double-digit returns.

This year, a portfolio would have required to be heavily weighted in American stocks to keep up with the S&P 500. However, having a high concentration of U.S. stocks would have resulted in a 55% portfolio decline during the financial crisis. If you find yourself tempted to concentrate only on the near-term gains in U.S. stocks, consider whether you would have been at ease with such a steep loss during the financial crisis. Short-term performance chasing and market timing are frequently caused by placing an excessive amount of emphasis on the best-performing asset class, which typically produces unsatisfactory effects for investors.

The goal of diversification is to reduce volatility. You can develop a strategy that works towards your long-term financial goals while minimizing the short-term ups and downs of investing by investing in a diversified portfolio that consists of a variety of various asset types.

For instance, each investment in a portfolio has a specific function:

Bonds can give income and are often less volatile than stocks. Commodities and real estate can help provide potential inflation protection. Cash provides ballast during the inevitable periods of turbulence. Stocks can bring growth over time but come with the potential for significant volatility.

A diverse portfolio that fits your risk tolerance may occasionally outperform the Dow or S&P 500 and underperform at other times. More aggressive portfolios that prioritized equities have typically performed well in previous months; but, more conservative portfolios that prioritized bonds performed better in the first quarter of 2018. To achieve your financial goals over time and be able to sleep at night, diversification entails putting together a portfolio of investments with a variety of traits.

The Obstacles We Face and How We Strive to Surmount Them

Even if the aforementioned is accurate, pretending that problems with having globally distributed teams don't exist would be unjustified. The obstacles to success as a worldwide team in a global market are real, and some of them are listed below along with suggestions for removing them.

Small Obstacles

There are many little things that are simple to overlook until you are actually faced with them. For instance, there are all of the following factors to take into account for something as routine as meetings or standups:

- What audiovisual equipment should you use for meetings?
- When is the optimum time of day to hold a remote meeting?
- Who should chair meetings, and should moderators be switched between locations?

Meetings present extra issues and difficulties to resolve, but so can seemingly little issues like the time of day in each office and how various offices observe (or don't observe!) Daylight Savings Time (DST). This has an effect on scheduling since it causes various degrees of working day overlap throughout the year.

Planning is difficult as well. Every nation has its own PTO/leave policy, holidays observed, sick leave regulations, etc. Our tech teams frequently use the scrum technique. Planning and forecasting work for upcoming sprints efficiently involves a thorough awareness of schedules for who will be in the office as well as knowledge of which offices have upcoming holidays. Where do we start? At the moment, we plan in a fairly traditional manner: We request that out-of-office notices be posted to a shared team calendar, and team members utilize scrums as a chance to let others know in general if they will be out of the office for a significant amount of time.

More Difficult Problems

The difficulties listed above might be considered minor, albeit perhaps not because considerable thought is needed to solve them. We refer to these issues as "little problems" instead because they require consideration and resolution in order to be handled by a globally dispersed team. However, there are other more covert and underlying difficulties in working on geographically dispersed teams. The more complicated issues usually result from misunderstandings, miscommunications, preconceptions, etc. that are cultural in nature.

We operate inside a framework and perspective that are inevitably influenced by our native culture. This happens to be so commonplace that it's simple to assume (incorrectly) that everyone thinks, behaves, and communicates in the same way. It can be simple to make assumptions or perceptions about people all around the world without stopping to think or notice that there might be cultural variations at work. This has the potential to negatively affect our capacity to collaborate productively and amicably across workplaces, which can be quite destructive.

Developing a Common Framework

Different conceptualizations of cultural differences exist. The "Culture Map" structure proposed by Erin Meyer in her book, The Culture Map, is the one that we've discovered to be useful. There are finally eight scales that sum up the many cultural disparities seen when working with people from diverse cultural origins, as described in this book:

• Low-Context (Explicit) vs. High-Context Communication (Implicit)
• Leading: Egalitarian vs. Hierarchical • Deciding: Consensual vs. Top-Down • Trusting:

Task-based vs. Relationship-based • Differing: Confrontational vs. Confrontation-Avoiding • Scheduling: Structured (Linear) Time vs. Flexible Time • Convincing: Deductive vs. Inductive Reasoning (or: Concept-First vs. Applications-First Reasoning)

Does a Common Framework Address All Issues? No. It's still simple to believe that all offices operate in the same manner in terms of how we think and work. But it is a beginning. The framework enables us to step back from the idea that "I am always correct." It has also highlighted a few significant "rules" that we would advise adhering to if you are a member of a project team that is geographically dispersed:

• Cultural differences exist and shouldn't be disregarded.

- Every society deviates from its own cultural norms. When working in a global team, what is "correct" within one culture could be "wrong." Therefore, the golden rule is to never presume that your method of thinking and doing is the best.
- Rather than expecting others to understand you, try to understand them. Alternatively, try to understand rather than trying to be understood. Both will occur, but if being understood is your main concern, you'll never truly understand other coworkers and you'll be operating under the false presumption that your cultural tastes are the best, wisest, etc.

Every cultural perspective has something to offer and can give you and your tech team a distinct advantage. Who really wants plain vanilla ice cream all the time, after all? Not us.

CHAPTER 14

Creating a Mega Project Culture

The study of megaprojects has received growing attention since the late 1980s (e.g.

Morris and Hough 1987). Cause-effect analysis concentrating on high-cost/high-risk analysis of economic efficiency and managerial performance has dominated research of knowing-doing gaps within megaprojects in the construction sector (cf. Stinchcombe and Heimer 1985; Flyvberg et al. 2003; Flyvberg 2005; Miller and Lessard 2000; Olds 2001). Merrow has taken a similar stance in the context of industrial megaprojects (2011). Merrow proposed a change in emphasis toward greater front-end planning and management using the lens of business objectives and the shaping process (cf. Miller and Lessard 2000). This gives rise to the viewpoint that a megaproject is the result of its own fragility. Therefore, ensuring that the megaproject is delivered safely throughout its lifecycle depends on establishing the mechanisms of causal variety and the contextual elements that define the megaproject environment. This paper focuses on the issues of culture and coordination (e.g. Berggren et al. 2001) in megaproject planning and management due to the fragility of megaprojects and the number of interfaces. When organizational fragmentation emerges, integration requires methodical management. However, in practice, overly predetermined integration leads to muddled thought and confusing action. The answer to the theoretical question "Is there a new route forward to underlying culture in the dynamic ecology of building Megaprojects beyond the static perceptions that dominate theories and practices?" is provided by this. To improve the ability to measure cultural dynamics in a megaproject temporary multi-organizational team, the goal is to implement a 4-class system (TMO). In this way, culture can come to be in order to better coordinate the TMO's performance over the course of the project lifecycle and make it more transportable, optimized, and effective.

Megaprojects, coordination, and culture

First, it's critical to stress that a megaproject is defined as a single project firm that manages a portfolio of projects under the temporary leadership of a coalition. This is further complicated by the fact that numerous TMOs are in charge of various projects within the program. Megaproject TMOs are complicated since their structure consists of numerous interconnected components. Its operation is distinguished by differentiation and interdependence due to the involvement of numerous distinct organizations (Baccarini 1996). Therefore, it's crucial to comprehend how culture supports this dynamic coordinating mechanism. In the past, research on construction project management has primarily emphasized increasing project efficiency rather than effectiveness (cf. Shenhar et al. 1997). However, Morris et al. (2011) warned against categorization of project management as a "execution-only oriented discipline." In other words, ignoring the socio-cultural system that is ingrained inside the company and project itself and depending solely on Weberian theories of precise application of skills, knowledge, tools, and techniques The notion that "the ideology and cultural prejudices of epistemic communities impact project processes in ways that favor displacement" is supported by Gellert and Lynch (2003). In order to manage projects effectively, the conventional cause-effect efficiency analysis needs to be augmented with knowledge of the institutional framework. Miller and Lessard's (2000) research also showed that efficiency and effectiveness should not be equated and that megaproject TMOs have a harder time achieving effectiveness due to ambiguous perceptions of the project and a lack of understanding of - and consequently, management of - project externalities. This section starts with a brief overview of previous research on culture in the context of megaprojects, then evaluates the dynamic aspects of culture, and concludes with a theoretical position on how measurements and dimensions of culture should be treated in megaprojects.

The Relationship in Megaprojects Between Culture and Coordination
Cultural differences at the country level are one of the elements that influence organizational effectiveness and, to a lesser extent, those at the

organizational level, according to project management researchers (e.g., Pant et al. 1996; Chen and Partington 2004; Phua and Rowinson 2003; Bredillet et al. 2010). In other words, management and organizational analysis cannot be isolated from the social paradigm that has emerged around the idea of culture. Of course, one could counter that the effects of cultural differences on projects and megaprojects may not be as obvious and obvious in terms of their immediate and obvious consequences as other difficulties. Others contend that the ultimate secret to project success is the use of established methods (cf. Milosevic and Patanakul 2005). The prerequisites that "automatically lead to enhanced project success" are not met when project management is reduced to a set of tools, procedures, and higher levels of standardization (Milosevic and Patanakul 2005;

Levitt 2011). Moving ahead, the context of megaprojects is distinct and original because it includes participants from various organizational and institutional settings who have dissimilar interests and priorities. The "logic of action" (Bresnen et al. 2004) is autonomous in this "complex system of interest groups, some congruent, some conflicting" (Cherns and Bryant 1984). Consequently, a megaproject coalition suggests rigorous integration across many public and commercial institutions responsible for the planning, design, and execution of individual projects within the program. Megaproject coalitions embody uncertainty and ambiguity, novelty, and complex systems. This highlights how difficult it is for the teams managing operations for construction projects, or TMOs for megaprojects, to coordinate. Only a small number of existing literatures have perhaps precisely examined how culture affects coordination in the context of megaprojects. Winch et al. (1997) released an essay based on the Transmanche Link, one of the initial efforts in this field, in the late 1990s. The authors of this study attempted to employ Hofstede's five national culture aspects as a theoretical framework to evaluate cross-border management at the organizational level. The researchers discovered that both at the operational project level and at the strategic program level, Hofstede's measurements did not accurately capture what actually occurred. In actuality, the five original dimensions' scores were the exact opposite of what happened in reality. As a result, a general conclusion was reached that the relationship between culture and coordination in megaprojects exists, poses a problem, and necessitates a deeper comprehension of the interactions between the business world, organizations, and national cultures. Therefore, it can be claimed that cultural diversity between countries is merely a general indicator of a successful collaborative effort. Early in the new millennium, Clegg et al. (2002) discovered that a big infrastructure project's culture is seen as a governance instrument. As a result, top-down efforts are made to actively foster a project-wide culture that places a strong emphasis on coalition-wide awareness and solidarity. The study's emphasis, however, seems to imply that culture is something that can be controlled scientifically. so ignoring the concept's contextual and dynamic components. In this way, value-based exchanges at relational and organizational interfaces are condensed to a single layer of contact. Similar studies of culture in the context of megaprojects have also been conducted by Van Marrewijk and his colleagues (e.g. van

Marrewijk 2005; van Marrewijk 2007; van Marrewijk et al. 2008; Smits and van Marrewijk 2012). The authors used a dynamic process method based on Martin's (2002) paradigm for anthropological research to examine how culture affects project coordination and success. These studies are founded on a single case study that examines several stages in time. The interdependence of various organizational components and the ramifications resulting from the larger institutional environment, however, are simplified. Additionally, the authors have not yet offered a comprehensive theoretical framework for how to proceed with identifying, operationalizing, and assessing crucial cultural components in the context of dynamic megaprojects. As a result, the richness of the data itself seems to eclipse the levels of analysis, making it difficult to distinguish between implications arising from a country's or an organization's level of culture. Table 1 provides a general overview of the study findings that have been published so far on the relationship between culture and coordination in massive building projects. In conclusion, it is not a stretch to argue that it is difficult to foresee the upfront form and size of effect resulting from the dynamics of cultural variety given current thinking in managing and comprehending the implications of culture in megaprojects. This leads us to the following subsection.

Culture as Dynamic: Culture has been seen as a crucial organizational concept in the construction industry, despite being vast and complex (cf. Tijhuis and Fellows 2011). The idea is ingrained in organizational studies and includes aligned behavior and coordinated action in addition to shared values and beliefs. Since the publication of Hofstede's (1980) seminal study on the five dimensions of national cultural differences, culture at the organizational level has received more attention and has been developed using anthropological and sociological perspectives to something that can help improve organizational and individual performance (e.g. Douglas 1999). Therefore, culture acts as a key component and benchmark to attain effectiveness and coherence between the various business units within and between enterprises. Additionally, academics have noted how culture is dynamic and multifaceted, changing gradually depending on the circumstances. However, project management literature has mostly distinguished between cultures based on understandings that are primarily based on eastern and western ideas (cf. Pant et al. 1996;

Chen and Partington 2004; Phua and Rowinson 2003; Bredillet et al. 2010). It is clear from the overview in Table 1 that attempts to regulate culture and universal, reductive, and static ideas of it are ineffective. French (2007) proposed five major cultural layers: I Global - the primary influence on behavior; ii National - the primary influence on attitudes; iii Regional - the primary influence on beliefs (within an ethnic group); iv Community - the primary influence on values (within an organization, group, or team); and v Personal - the primary influence on presumptions.

These layers interact with one another. According to a "spectrum of institutional or society-wide elements," culture is complicated and multifaceted at every given layer or level (French 2007). This complex aspect of culture creates dynamic interactions between top-down and bottom-up value-based interactions.

These dynamic value-based exchanges in a megaproject are primarily caused by: I Diverse stakeholder influence and constraints (different stakeholder influence at different points in the project lifecycle); ii) Variation in socio-cultural contexts affecting operation; and iii) International stages with geopolitical interests. Particularly with regard to high-profile undertakings that entail numerous subjective roles and political critics, iv Differentiation of units and specialization, v Temporary - Has a start and end. However, the coalition's members have somewhere to go before and after the lifecycle time. This is due to the coalition's vi decentralized command systems with dominating coalition and vii reciprocal interdependency between the contracting organizations.

To establish knowledge of consistent recipes within the project coalition so that other people can complete their part of the task without constant supervision, an effective megaproject culture should be redefined. This will create a precept for actions as well as a scheme of expression and a scheme of interpretation. This perspective and method of approaching the issue focuses on and addresses how various institutional levels of culture are assimilated, become the accepted organizational culture, and further involve and influence the mechanisms, processes, and structuring of decision-making at the organizational and project levels.

To begin with, this book is intended to be a literary reflection of the cultural studies that have been and should continue to be prioritized in the context of megaprojects. It provided an evolutionary viewpoint by viewing the administration of megaprojects as a kaleidoscope of transient organizational units

(TMOs). The old Hofstede-inspired and other conventional values studies that assess the effects of organizational culture based on static right and wrong or coherence and incoherence are made difficult by the third wave idea. Existing literatures have different levels of analysis from one another. On the one hand, organizational level cultural disparities are explained using data collected at the national level. However, these levels are mixed together, and it was asserted that culture could be governed scientifically from the top-down while ignoring the bottom-up opposition.

In order to understand how culture changes over time within megaproject TMOs, this book attempted to blend cultural dynamics, ecology, and development. Given the complicated and transient setting of a megaproject TMO coalition, linkage between cultures at the national, institutional, organizational, and project levels needs to be isolated and determined. To strengthen the ability to identify propensities and predictions for a more transferrable, optimized, and effective continuity of the TMO's performance across the project lifespan, new theoretical starting points are thus provided. It is demonstrated that culture persists as an ecosystem of symbiotic inter-organizational relationships in a diversified megaproject setting. Relational interfaces faced by the TMO as a social construct provide limits in the dynamics and evolution of a successful culture by highlighting two primary coordination issues coming from this environment.

www.ingramcontent.com/pod-product-compliance
Lightning Source LLC
Chambersburg PA
CBHW070319220526
45465CB00013B/995